W0245953

148
Advances in Polymer Science

Springer-Verlag Berlin Heidelberg GmbH

Interfaces
Crystallization
Viscoelasticity

With contributions by
A. Budkowski, I. W. Hamley,
T. Koike

 Springer

This series presents critical reviews of the present and future trends in polymer and biopolymer science including chemistry, physical chemistry, physics and materials science. It is addressed to all scientists at universities and in industry who wish to keep abreast of advances in the topics covered.

As a rule, contributions are specially commissioned. The editors and publishers will, however, always be pleased to receive suggestions and supplementary information. Papers are accepted for „Advances in Polymer Science" in English.

In references Advances in Polymer Science is abbreviated Adv. Polym. Sci. and is cited as a journal.

Springer WWW home page: http://www.springer.de

ISSN 0065-3195

ISBN 978-3-662-15629-2   ISBN 978-3-540-48836-1 (eBook)
DOI 10.1007/978-3-540-48836-1

Library of Congress Catalog Card Number 61642

© Springer-Verlag Berlin Heidelberg 1999
Originally published by Springer-Verlag Berlin Heidelberg New York in 1999
Softcover reprint of the hardcover 1st edition 1999
The use of registered names, trademarks, etc. in this publication does not imply, even in the absence of a specific statement, that such names are exempt from the relevant protective laws and regulations and therefore free for general use.

Typesetting: Data conversion by MEDIO, Berlin
Cover: E. Kirchner, Heidelberg
SPIN: 10702230   02/3020 - 5 4 3 2 1 0 - Printed on acid-free paper

Editorial Board

Contents

Interfacial Phenomena in Thin Polymer Films: Phase Coexistence and Segregation

Andrzej Budkowski

Institute of Physics, Jagellonian University, Reymonta 4, PL-30–059 Kraków, Poland
E-mail: *ufbudkow@cyf-kr.edu.pl*

Interfaces formed or exposed by polymer mixtures are often employed in modern techno-logical applications. This is especially true for thin polymer films used nowadays in pho-toresist lithography, electrooptical devices or nanometer-scale surface patterning. Apart from these technology-oriented aspects the interfacial phenomena in thin polymer films pose a fundamental scientific challenge to physics of polymers and thermodynamics of condensed matter. This work reviews experimental results on the equilibrium properties of interfaces created by polymer mixtures confined in thin films. It confronts experimental data with theoretical expectations based mainly on mean field models. Most of the data have been obtained by high resolution profiling techniques emerged in the last decade. These techniques allow us to trace concentration vs depth profiles across a thin film with a depth resolution better than the characteristic size of a polymer coil. The interfacial phe-nomena of phase coexistence and segregation are described as observed in thin polymer films. This work also considers related issues in the focus of current research such as wet-ting phenomena, finite size effects expected in very thin films, surface (and bulk) properties of mixtures with stiffness disparity and conformational properties of end-segregated mac-romolecules forming polymer brushes.

Advances in Polymer Science, Vol.148
© Springer-Verlag Berlin Heidelberg 1999

List of Symbols and Abbreviations

The numbers in brackets "{ }" are the section numbers where the symbol or the abbreviation is first defined or reintroduced.

α_i	segmental polarizability {2.2.2}
β	SCMF adsorption parameter {4.1}
β'	(mean field) adsorption parameter $\beta'=N_A\chi_{AP}-\Delta_i$ {4.1}
β'_{SCMF}	β' calculated from β for assumed w_A {4.2.3}
γ_{ij}	interfacial tension (i, j=1 (ϕ_1), 2 (ϕ_2), s (surface)) {3.1}
Γ	aspect ratio $\Gamma=a/V^{1/3}$ {3.1.2.2}
δ, δ_i	solubility parameter of species i {2.2.2}
$\delta_{ei,xi}$	solubility parameter of deuterated (e_i) random copolymer x_i {2.2.3}
$\delta_{0,xi}$	solubility parameter of hydrogenous random copolymer x_i {2.2.3}
$\delta_h-\delta_d$	difference of δ of protonated and deuterated copolymer ($\delta_{0,xi}-\delta_{1,xi}$) {2.2.3}
$\delta_0-\delta_1$	local $\delta(x)$ slope {2.2.3}
$\Delta\alpha$	difference of segmental polarizabilities of two species {2.2.2}
$\Delta\gamma$	surface tension difference of pure blend components {3.1.2.2}
$\Delta\mu$	difference of chemical potentials of species ($\mu_A-\mu_B$) {2.1}
ΔA	difference of Hamaker constant of species {3.1.2.1}
$\Delta f(\phi; \chi, \phi_\infty)$	local excess energy for reservoir concentration ϕ_∞ {2.1}
$\Delta f(\phi;\chi,\Delta\mu,\phi_b)$	local excess energy for reservoir concentration ϕ_b {3.2.1}
ΔF_M	Gibbs free energy of mixing per lattice site {2.1}
ΔF_e	difference between F_e of partial- and complete-wetting {3.1.1}
Δf_s	difference between f_s of two pure blend components {3.1.2.2}
Δ_i	energy reduction due to presence of block A of A-N at interface {4.1}
ε	surface critical exponent of surface tension difference ($\gamma_{1s}-\gamma_{2s}$) {3.1}
ε_{ij}	contact energy between i- and j-segments {2.1}
Θ	contact angle of two coexisting phases at surface {3.1}
$\kappa(\phi)$	coefficient of concentration gradient term {2.1}
$\Lambda(\phi)$	mutual mobility {2.2.1}
μ, μ_A, μ_B	chemical potential of polymer species A (B) {2.1}
μ_1	linear term coefficient in the expression for $f_s(\phi_s)$ {3.1.1}
μ_{brush}	chemical potential of copolymer chains in brush layer {4.1}
$\mu^{L(S)}_{brush}$	chemical potential of longer (shorter) chains in mixed brush layer {4.2.4}
μ_{bulk}	chemical potential of copolymer chains in bulk region {4.1}
$<\mu_{bulk}>$	adjusted bulk chemical potential of copolymer chains {4.1}
$\mu^{L(S)}_{bulk}$	chemical potential of longer (shorter) chains in bulk region of film {4.2.4}
ν	bulk critical exponent of interfacial tension γ_{12} {3.1}
ξ	correlation length {2.1}

ξ_b	correlation length at binodal $\xi_b=\xi(\phi_1)=\xi(\phi_2)$ {2.1}
$\xi_{\|}$	transverse correlation length {3.2.2}
ρ	depth resolution of profiling techniques (HWHM of Gaussian function) {1}
ρ_L	lateral spatial resolution {1}
σ	surface coverage by brush N-mer chains $\sigma=a^2/\Sigma$ {4.1, 4.2.1}
σ^L, σ^S	coverage by longer (shorter) chains of mixed brush {4.2.4}
$\sigma_1,...,\sigma_5$	coverage values for cross over between different $L(\sigma)$ regimes {4.1}
Σ	mean area per chain comprising brush layer {4.1}
$\phi(z)$	local volume fraction of blend component A at depth z {2.1}
ϕ_b	local composition with plateau in $\phi(z)$ profile {3.2.1}
ϕ_c	critical volume fraction {2.1}
ϕ_c^{D}	critical volume fraction of very thin film between symmetric walls {3.2.2}
ϕ_{cmc}	critical copolymer concentration for onset of micelle formation {4.2.3}
ϕ_{fw}	volume fraction of "frozen wall" created by immobilized polymer blend {3.1.2.2}
$\phi_s, \phi_{si}, \phi_{is}$	surface volume fraction of species A (its specific values, i=1,2,d,e) {3.1}
$\phi_{IL(R)}(z,t)$	composition profile at left (right) side of interface I(t) {2.2.1}
ϕ_s^L, ϕ_s^R	vol. fraction at left/right "external" interface bounding thin polymer film {3.2.1}
ϕ^L, ϕ^S	bulk composition of long (short) diblock copolymers {4.2.4}
ϕ_1, ϕ_2	volume fractions of coexisting phases {2.1}
ϕ_∞	bulk composition of thin film considered as semi-infinite mixture {3.1}
$\nabla\phi$	volume fraction gradient along z axis {2.1}
Φ	composition of N-mer brush chains in brush layer {4.1}
$\chi, \chi(\phi,T)$	Flory-Huggins interaction parameter {2.1, 2.2.2}
χ_{AP}	interaction parameter between species of "anchor" block A and matrix P-mer {4.1}
χ_{AN}	interaction parameter between species of blocks A and N of copolymer A-N {4.1}
χ_{NP}	interaction parameter between species N of A-N and matrix P-mer $\chi_{NP}=\chi$ {4.1}
χ_c	critical interaction parameter $\chi(T_c)$ {2.1}
$\chi_{E/EE}$	microstructural contribution to $\chi(hx_j/dx_i,e_i)$ {2.2.3}
$\chi_{h/d}$	isotopic contribution to $\chi(hx_j/dx_i,e_i)$ {2.2.3}
χ_s	surface energy difference parameter {3.1.2.2}
χ_{SX}	interaction parameter between surface and species X {3.1.2.2}
χ_{SANS}	effective interaction parameter as determined by SANS {2.1}
$\chi(hx_j/dx_i,e_i)$	interaction parameter between protonated x_j and deuterated (to extent e_i) x_i {2.2.3}

Ω	lattice site volume {2.1}
I, A/B	binary mixture composed of polymers A and B {2.2.1}
II, A//B	interface between phases rich in A and B {2.2.1}
a, a_k	statistical segment length of polymer k (k=A, B, x, i, j) {2.1}
A	area of interface {2.1}
A-N	diblock copolymers with N_A "anchor" segments A and N "buoy" segments {4.1}
AFM	atomic force microscopy {1}
(COOH)dPS	carboxy terminated deuterated polystyrene {4.2.1}
D_0	specific value of diffusion coefficient {2.2.1}
D	overall thin film thickness {3.2.1}
dPS	deuterated polystyrene {2.2.1}
dPS-PVP	diblock copolymers composed of dPS and PVP blocks {4.2.3}
dL, dS	deuterated long (short) component of bimodal diblock mixture {4.2.4}
dx (e.g., d75)	partly deuterated random copolymer $((C_4H_8)_{1-x}(C_2H_3(C_2H_5))_x)_N$ {2.2.1}
$(-df_s/d\phi)_s$	compositional derivative of "bare" surface free energy {3.1.1}
e, e_k	fractional deuteration of deuterated species k (k=1, 2, i (x_i), j (x_j)) {2.2.3}
E_{int}	internal (short-ranged) energy of polymer blend in bulk {3.1.2.2}
$E_{1-x}EE_x$	random olefinic copolymer composed of E=C_4H_8 and EE= $C_2H_3(C_2H_5)$ {2.2.1}
f_s	(short- ranged) "bare" surface free energy {3.1.1, 3.2.1}
f_s^H	enthalpic contribution to "bare" surface free energy {3.1.2.2}
$f_s^{H,mn}$	contribution to f_s due to missing neighbor effect {3.1.2.2}
f_s^L	f_s term due to left interface bounding thin polymer film at z=0 {3.2.1}
f_s^S	entropic contribution to "bare" surface free energy {3.1.2.2}
f_s^R	f_s term due to right interface bounding thin polymer film at z=D {3.2.1}
$F(\phi(z))$	functional F_b with "bare" surface free energy terms {3.1.1, 3.2.1}
$F_b(\phi(z))$	functional of free energy of mixing {2.1}
F_e	total excess free energy of equilibrium profile $\phi(z)$ {2.1, 3.1.1}
FH	Flory-Huggins lattice model {2.1}
FRES	forward recoil spectrometry {1}
F_{tot}	overall free energy of mixed brush layer {4.2.4}
g	quadratic term coefficient in expression of $f_s(\phi_s)$ {3.1.1}
G	free energy per one brush chain in brush layer {4.1}
G^L, G^S	free energy per one longer (shorter) chain of mixed brush layer {4.2.4}
H	offset value of z characteristic for the real brush profile {4.2.1}
hS	protonated short component of binary diblock mixture {4.2.4}
hx (e.g., h75)	random olefinic copolymer $((C_4H_8)_{1-x}(C_2H_3(C_2H_5))_x)_N$ {2.2.1}
HWHM	half width at half maximum {1}

$I(t)$	depth z position of "internal" interface at time t {2.2.1}
$I_e(x,y)$	depth z position of "internal" interface between coexisting phases {2.2.2, 3.2.2}
J	flux of polymer species A {2.2.1}
k_B	Bolzmann constant
$L, L(\sigma)$	average height of brush layer {4.1}
LCST	lower critical solution temperature T_c {2.2.2}
m.w.	molecular weight {2.2.1}
N, N_x, N_i	polymerization index of polymer (copolymer block) i=A, B, N ($N=N_N$) {2.1}
N_c	polymerization index of whole diblock copolymer $N_c=N_A+N$ {4.1}
N^L, N^S	polymerization index of longer (shorter) copolymer of bimodal mixture {4.2.4}
NR	neutron reflectometry {1}
NRA	nuclear reaction analysis {1}
P	(polymerization index of) matrix homopolymer molecules {4.1}
PBD	polybutadiene {4.2.1}
PBr_xS	partially brominated polystyrene {3.1.2.1}
PEP	poly(ethylene propylene) {3.1.2.1}
PI	polyisoprene {4.2.2}
PI-dPS	diblock copolymers composed of polyisoperene (PI) and dPS blocks {4.2.1}
PI-PS	diblock copolymers composed of polyisoperene (PI) and PS blocks {4.2.2, 4.2.4}
PMMA	poly(methyl methacrylate) {3.2.2, 4.2.1}
PPO	poly(2,6-dimethylphenylene oxide) {4.2.1}
PS, hPS, dPS	polystyrene (protonated, deuterated) {2.2.1}
PS-PMMA	diblock copolymers composed of PS and PMMA {4.2.2}
PVME	poly(vinylmethyl ether) {3.2.2, 4.2.1}
PVP	poly(2-vinylpyridine) {4.2.2}
PVT data	thermal expansion coefficient and isothermal compressibility data {2.2.3}
q	scattering vector {2.1}
q	exponent of brush conformation relation $L \propto \sigma^q$ {4.1}
q_m	wave number of dominant mode of early stage spinodal decomposition {2.2.1}
Q, Q_e, Q_d	specific surface volume fraction value at which $(-df_s/d\phi)_s$ changes its sign {3.1, 3.1.1}
r_c	asymmetry ratio of diblock copolymer $r_c=N_A/N_c=N_A/(N+N_A)$ {4.2.3}
$R_g, R_g(N)$	radius of gyration of polymer coils $R_g=(N/6)^{1/2}a$ {2.1}
$R_0, R_0(N)$	mean square end-to-end distance of polymer coils $R_0=N^{1/2}a$ {4.1}
$R(t)$	time evolution of average domain size during spinodal decomposition {2.2.1}

$S(q)$	structure factor at scattering vector q {2.1}
SANS	small angle neutron scattering {1}
SCMF	self consistent mean field {3.1.2.1, 4.1}
SIMS	secondary ion mass spectroscopy {1}
SNOM	scanning near-field optical microscopy {1}
SSL	strong segregation limit {2.1}
t	time variable
T	absolute temperature
T_c	critical temperature {2.1}
T_c^D	critical temperature of a very thin film {3.2.2}
TEM	Transmission Electron Microscopy {1}
T_g	glass point temperature {2.2.1}
T_{pw}	pre-wetting critical point {3.1}
T_{ref}	reference temperature {2.2.3, 3.1.2.3, 3.1.2.5}
T_w	wetting temperature {3.1}
$U(z)$	potential affecting "anchor" block A of copolymer A-N {4.1}
UCST	upper critical solution temperature T_c {2.2.2}
V	segmental volume {2.1}
$v(z)$	long-ranged contribution to surface free energy {3.1.2.1}
w	"intrinsic" interfacial width {2.1}
w_{exp}	experimental value of interfacial width {2.2.2}
w_A	width of interfacial region where "anchor" block A is confined {4.1}
w_D	effective interfacial width, $w_D \neq w$ due to capillary waves {3.2.2}
WSL	weak segregation limit {2.1}
$w_{1/2}$	interfacial half-width {2.1}
x, x_i (e.g., 75)	(composition of) random copolymer $((C_4H_8)_{1-x}(C_2H_3(C_2H_5))_x)_N$ {2.2.3}
z	axis perpendicular to "external" interface of a thin film
$z(\phi)$	reciprocal function of local composition ϕ vs depth z {3.1.2.1}
$z(\phi_\infty)$	distance from surface to plateau in $\phi(z)$ {3.1}
z_b	bulk (lattice) coordination number {2.1}
z_s	surface (lattice) coordination number {3.1.2.2}
z^*	surface excess {3.1, 3.1.2.1, 4.2.1}
$z^*_1(\phi_s)$	z^* for profile $\phi(z)$, starting at $\phi_\infty \to \phi_1$ and cut off at ϕ_s {3.1.2.3}
z^*_N	surface excess of N-mer brush forming blocks, $z^*_N = (1-r_c)z^*$ {4.2.3}
z^{*L}_N, z^{*S}_N	z^*_N of long (short) component of bimodal diblock mixture {4.2.4}

1

Introduction

Polymer blends have been used for decades [1] as relatively inexpensive materials with desirable structural and functional characteristics. Interfaces formed or exposed by polymer mixtures are often employed in modern technological applications. This is especially true for thin polymer films used nowadays [2] in photoresist lithography, electrooptical devices, gas separating membranes, or nanometer-scale surface patterning. The nature of the interface specifies [3] important properties such as adhesion, interfacial fracture toughness, friction and wear, adsorption, wettability, compatibility with adjacent phases, etc.

Apart from these technology-oriented aspects, the interfacial phenomena in thin polymer films pose a fundamental scientific challenge to physics of polymers [4, 5] and thermodynamics of condensed matter [6–9]. While some of the concepts necessary to describe these interfaces are typical for polymer melts, others (e.g., wetting) refer to liquids in general. Polymer mixtures are known to be ideal systems to test thermodynamic models based on a mean field approach. The mean field character is assured by the chain-like structure of polymers. Individual polymer coils are strongly entangled with each other, which coerces the average character of relevant segmental interactions [10]. In addition, the connectivity of the segments along the chain increases the characteristic scales in space [11, 12], time [13, 14], and temperature [15, 16] domains as compared to the situation in simple liquids. This allows easy and convenient observation of the interfacial phenomena.

This work reviews experimental results on the equilibrium properties of interfaces created by polymer mixtures confined in thin films. It confronts experimental data with theoretical expectations based mainly on mean field models. Some of these theoretical descriptions have been surveyed recently by Binder [6, 7].

Two types of interfaces are considered:
i) "Internal" interfaces between coexisting phases
ii) "External" interfaces separating a polymeric film from a vacuum or a substrate.

The *interfacial phenomena* of *phase coexistence* and *segregation* are described as observed in thin polymer films. This work also considers related issues in the focus of current research such as: *wetting phenomena, finite size effects* expected in very thin films, surface (and bulk) properties of *mixtures with stiffness disparity*, and conformational properties of end-segregated macromolecules forming *polymer brushes*.

Presented polymer mixtures are composed of amorphous macromolecules with different molecular architecture: homopolymers and random copolymers, with different segments distributed statistically along the chain, form partly miscible isotopic and isomeric model binary blends. The mixing of incompatible polymers is enforced by two different polymers covalently bonded forming diblock copolymers. Here only homopolymers admixed by copolymers are considered. The diblock copolymer melts have been described recently in a separate review by Krausch [17].

Most of the experimental data reviewed in this work have been obtained by high resolution profiling techniques emerging in the last decade. These techniques allow us to trace concentration vs depth profiles depicting the local volume fraction of given polymer across an interface or a whole thin film layer [18] with a depth resolution better than the characteristic size of a polymer coil. This work is illustrated mainly by author's own results, obtained using the (non-resonant) nuclear reaction analysis (NRA) [19–21]. NRA is characterized by the depth resolution ρ of ca. 7 nm HWHM (Half Width at Half Maximum of the Gaussian function) at the vacuum//polymer interface deteriorating to $\rho \approx 30$ nm HWHM at a depth of about 600 nm. We also quote here the data of other real space- and momentum space-profiling techniques: the first group consists of forward recoil spectrometry (FRES) [22, 23] and secondary ion mass spectroscopy (SIMS) [24–27] while the second one includes X-ray and neutron reflectivity (XR, NR) [28]. The profiles yielded by the real space-techniques are straightforward but do not resolve very small scale details (less than a few nm) due to a limited depth resolution. On the contrary, the profiles obtained by reflectivity methods are model dependent although they are characterized by an excellent resolution of 1 nm HWHM. Comprehensive comparison of profiling techniques is given by Kramer [29].

This work also considers other data relevant for polymer mixtures in bulk and in thin films. The coexistence conditions in the bulk are evaluated by small angle neutron scattering (SANS) [30, 31]. Complex phase domain morphologies are discussed as determined in thin films, using quasi three-dimensional profiling. It combines the analysis of the profile, depicting the average composition normal to the "external" interfaces, with the examination of lateral structures observed by microscopic methods. The latter are optical microscopy (lateral resolution ρ_L of about 0.5 µm) [32], scanning near-field optical microscopy (SNOM, $\rho_L \approx 100$ nm) [33], transmission electron microscopy (TEM, $\rho_L \approx$ a few nanometers) [34], or atomic force microscopy (AFM which allows studies on a nanometer scale) [35].

This review is organized as follows.

In Sect. 2 the coexistence conditions of high polymer mixtures are described. Here we focus on the "internal" interface between two coexisting phases with a bilayer morphology. The properties of this interface determine phase coexistence characteristics necessary to describe segregation phenomena discussed in Sects. 3 and 4.

Section 3.1 considers the segregation from binary polymer blends towards "external" interface of a thin film described in a semi-infinite mixture approach. We relate the segregation with wetting phenomena. The role of a vapor and a gas in a classic formulation of this problem is played by two coexisting polymer phases.

In Section 3.2 both "external" interfaces confining binary polymer mixture in a thin film geometry are explicitly considered. These interfaces specify the equilibrium morphologies of the coexisting phases. Finite size effects relevant for thin films with reduced thickness are also described.

Section 4 focuses on the segregation of diblock copolymers admixed to homopolymer matrices. The copolymers form brush-like layers composed of chains attached by their "anchor" blocks to the "external" interface. Similar dou-

ble brush layers are formed at the "internal" interface. The conformation properties of polymer brushes are discussed and related to the segregation properties
of the diblocks.

Each section, from Sect. 2 to Sect. 4, starts with a brief introduction into relevant description within the mean field approach, used later to analyze the reviewed experimental results, and ends with a summary and conclusions.

2
Phase Coexistence in Binary Polymer Blends

2.1
Mean Field Theory

The thermodynamic properties of binary polymer mixtures are described by the
Flory-Huggins (FH) lattice model [36, 37]. In this approach each lattice site is
occupied by one effective segment of the polymer A or the polymer B. On the
considered lattice polymer chains, with degree of polymerization N_A or N_B, are
represented as random walks of N_A or N_B steps, respectively, with assumed
Gaussian distribution of distances between their ends. For given volume fractions, $\phi_A = \phi$ and $\phi_B = 1 - \phi$, of both polymers (defined as numbers of lattice sites
taken by A or B chains, normalized with the total number of the lattice sites) the
(Gibbs) free energy of mixing per lattice site ΔF_M is given by the following expression [4, 38]:

$$\left.\frac{\Delta F_M}{k_B T}\right|_{site} = \frac{\phi}{N_A}\ln\phi + \frac{1-\phi}{N_B}\ln(1-\phi) + \chi\phi(1-\phi) \tag{1}$$

The first two terms in Eq. (1) account for the combinatorial entropy of mixing. They are analogous to those present for simple liquids, except of the $1/N_A$
and $1/N_B$ factors originating from the connectivity of the chains. The third term
represents in the mean field approximation the interaction between segments A
and B occupying neighboring lattice sites (any correlations in the occupancy of
neighboring sites are neglected). In the original formulation the FH interaction
parameter χ is expressed by the contact energies ε_{ij} between the segments i and j:

$$\chi = \frac{z_b - 2}{k_B T}[\varepsilon_{AB} - \frac{1}{2}(\varepsilon_{AA} + \varepsilon_{BB})] \tag{2}$$

where z_b is the lattice (bulk) coordination number. The $(z_b - 2)$ term accounts for
the fact that, for each lattice site, two neighboring sites must be occupied by segments of the same chain. Similar effects are, however, neglected when entropy of
mixing terms of Eq. (1) are evaluated. In addition, effects due to chains ends are
disregarded throughout.

The most drastic approximations leading to Eq. (1) are credited by the following arguments [39]: a chain conformation of a polymer coil in dense melt sys-

tems, as described here, is Gaussian; further, the Ginzburg criterion [40] (the condition for cross-over from the mean field to Ising-type behavior) can be written down as $|1-\chi_c/\chi| \sim 1/N$ [7, 10], where χ_c is the χ value corresponding to critical temperature T_c. Thus for high polymers, i.e., for those with $N_A \approx N \approx N_B$ of the order of 10^3–10^4, the non-mean field region shrinks to a small area around the critical point [41].

The validity of the other approximations in the treatment of the lattice model is still an open question. Also the disparity in size and/or shape, as well as the compressibility of the real blend components, present even for the nearly ideal case of isotopic mixtures, makes the FH lattice model slightly artificial. In spite of this criticism, Eq. (1) is commonly used for fitting experimental phase diagrams and scattering data as no accepted expression replacing it has yet emerged [39, 42]. As a result the effective interaction parameter χ, obtained by the fitting procedure, often depends on ϕ and N. The explanation of the physical nature of the effects leading to such a complicated nature of the effective χ parameter is the objective of many theoretical works [39, 42–48].

On the basis of standard criteria for equilibrium, stability limits, and criticality yielding coexistence curve (binodal), spinodal line, and critical point, the phase behavior may be predicted using Eq. (1):

binodal:
$$\frac{\Delta F_M(\phi_2) - \Delta F_M(\phi_1)}{\phi_2 - \phi_1} = \left.\frac{\partial \Delta F_M}{\partial \phi}\right|_{\phi_1} = \left.\frac{\partial \Delta F_M}{\partial \phi}\right|_{\phi_2} \tag{3a}$$

spinodal:
$$\frac{\partial^2 \Delta F_M}{\partial \phi^2} = 0 \tag{3b}$$

critical point:
$$\frac{\partial^3 \Delta F_M}{\partial \phi^3} = 0 \tag{3c}$$

For a given temperature (i.e., for a given interaction parameter χ) the double-tangent construction of Eq. (3a) gives the volume fractions ϕ_1 and ϕ_2 of two co-existing phases (see Fig. 1). In the region of the phase diagram bounded by the coexistence curve a phase separated state of two phases ϕ_1 and ϕ_2 is the equilibrium stable state. The homogeneous mixture is stable outside the coexistence curve, metastable between binodal and spinodal lines (the "nucleation and growth" region), and unstable inside the spinodal line (the "spinodal decomposition" regime) [10, 49]. Equation (3), together with Eq. (1), are used to extract effective interaction parameter χ from coexistence curve and spinodal points data [9]. The shape of the phase diagram may be modified by the degree of polymerization of blend components. For instance for ϕ-independent χ parameter the combination of Eqs. (3b) and (3c) yields the following prediction for the location of the critical point: $\phi_c = N_B^{1/2}/(N_A^{1/2}+N_B^{1/2})$ and $\chi_c = (N_A^{1/2}+N_B^{1/2})^2/(2N_A N_B)$. For symmetric blends, i.e., $N_A = N_B = N$, FH lattice model yields $\chi_c = 2/N$ and predicts the critical temperature T_c to be proportional to N. This result was confirmed by

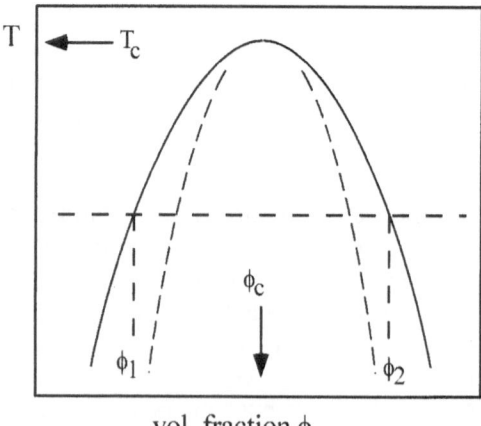

vol. fraction ϕ

Fig. 1. Temperature T vs volume fraction ϕ phase diagram of a binary polymer blend. *Solid line* denotes the coexistence curve (binodal) while the *fine dashed line* marks the spinodal line. Binodal connects with spinodal at the critical point (ϕ_c, T_c)

the experiment [50] and the computer simulation [51] (contrary to predictions of the other theory [52] competing with FH model).

Scattering techniques, such as Small Angle Neutron Scattering (SANS), probe the growth of the local volume fraction fluctuations close to the critical point T_c. The structure factor $S(q)$, describing the scattering intensity at a scattering vector q, is given by [4, 43, 49]:

$$S^{-1}(\vec{q}) = [\phi S_A(\vec{q})]^{-1} + [(1-\phi)S_B(\vec{q})]^{-1} - 2\chi_{SANS} \qquad (4)$$

The first two terms correspond to the combinatorial entropy terms of Eq. (1) and form the non-interacting part of the structure factor which is just a weighted average of the single-chain structure factors $S_A(q)$ and $S_B(q)$ of both blend components. $S_A(q)$ and $S_B(q)$ are characterized by the radius of gyration $R_{gA} = a_A(N_A/6)^{1/2}$ and $R_{gB} = a_B(N_B/6)^{1/2}$, where a_A and a_B are the statistical segment lengths of polymer A and polymer B, respectively. The last term of Eq. (4) yields the SANS determined interaction parameter χ_{SANS}:

$$\chi_{SANS} = -\frac{1}{2}\frac{\partial^2}{\partial\phi^2}[\chi(\phi)\phi(1-\phi)] \qquad (5)$$

which is equivalent to the χ parameter in Eq. (1) only if χ is ϕ-independent. In practice χ_{SANS} is determined from the structure factor S extrapolated to $q=0$, where it is simply related to the curvature of the free energy ΔF_M [4, 49]: $S^{-1}(0) = \dfrac{\partial^2 \Delta F_M}{\partial\phi^2}$. The correlation length ξ of concentration fluctuations is defined by the structure factor S at the $q \to 0$ limit: $S(q)=S(0)/(1+q^2\xi^2)$ and expressed by the following relation:

$$\xi^2 = \frac{a^2(\phi)}{36}[\frac{1-\phi}{2N_A}+\frac{\phi}{2N_B}-\phi(1-\phi)\chi_{SANS}]^{-1}.$$ (6)

The effective statistical segment length $a(\phi)$ is related to a_A and a_B: $a^2(\phi)=(1-\phi)a^2_A+\phi a^2_B$.

The structure of the interface formed by coexisting phases is well described by the Cahn-Hilliard approach [53] (developed in a slightly different context by Landau and Lifshitz [54]) extended to incompressible binary polymer mixtures by several authors [4, 49, 55, 56]. The central point of this approach is the free energy functional definition that describes two semi-infinite polymer phases: ϕ_1 and ϕ_2 separated by a planar interface (at depth z=0) and the composition $\phi(z)$ across this interface. The relevant functional F_b for the free energy of mixing per site volume Ω (taken as equal to the average segmental volume V of both blend components) and the area A of the interface is expressed by

$$\frac{F_b}{k_BTA/\Omega} = \int\limits_{-\infty}^{\infty}dz\{\Delta F_M[\phi(z)]-\Delta\mu\phi(z)+\kappa(\phi)[\nabla\phi(z)]^2\}$$ (7)

where ΔF_M is the mixing free energy given by Eq. (1), $\kappa(\phi)=a^2(\phi)/[36\phi(1-\phi)]$ and $\Delta\mu$ is the difference of the chemical potential $\Delta\mu=(\mu_A-\mu_B)=\partial\Delta F_M/\partial\phi(\phi_\infty)$. Here the bulk concentration ϕ_∞ is the volume fraction of the blend region homogeneous in composition (i.e., with $\nabla\phi=0$). The square gradient term describes the free energy costs of volume fraction inhomogeneities and is related to configurational entropy of Gaussian chains [7, 57]. The contributions to the non-local free energy in higher powers of the composition gradient may be obtained in rigorous calculations [58, 59] and they are proven to be irrelevant [58] for the weak segregation limit (WSL), i.e., for regions close to critical temperature T_c and small $(\chi-\chi_c)/\chi_c$ values. For the strong segregation limit (SSL) (large $(\chi-\chi_c)/\chi_c$ values) Roe [56] suggested the functional identical to this of Eq. (7) except that the coefficient $\kappa(\phi)$ is 1.5 times larger. It is important to stress that Eq. (7) is valid only for slow spatial variations of the volume fraction: $1/[\nabla\phi(z)]>aN^{1/2}$ (long wavelength limit: a is the average value of the statistical segment length).

The equilibrium solution of the problem is given by Euler-Lagrange equation applied to the functional F_b. At first, neglecting the gradient term, we obtain $\Delta\mu=\partial\Delta F_M/\partial\phi(\phi_1)$ corresponding to the bulk region of the homogeneous phase ϕ_1 ($\phi_\infty=\phi_1$). In general we obtain a second order differential equation which, integrated, yields

$$\kappa(\phi)[\nabla\phi]^2 = \Delta F_M(\phi)-\Delta F_M(\phi_1)-\Delta\mu(\phi-\phi_1):=\Delta f(\phi;\chi,\phi_1)$$ (8a)

We may also note an analogy between mean field theory and classical mechanics, and treat the integrand of the F_b functional as the Lagrangian: Then

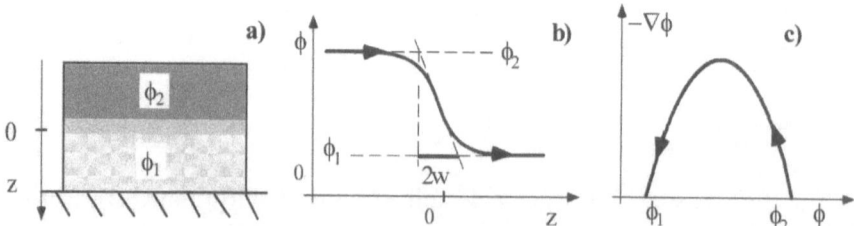

Fig. 2.a Polymer film composed of two coexisting phases ϕ_1 and ϕ_2. **b** The composition $\phi(z)$ profile across the interface with denoted interfacial region 2w. **c** Corresponding "phase portrait" $-\nabla\phi(\phi)$ relation (see Eqs. 10 and 11) as described by mean field approach

Eq. (8a) is obtained as the equation expressing kinetic energy [60]. For the bulk homogeneous ($\nabla\phi=0$) region of the second coexisting phase, $\phi=\phi_2$ and Eq. (8a) leads directly to a common-tangent criterion of Eq. (3a). The right-hand side of Eq. (8a) defines the local excess free energy $\Delta f(\phi; \chi, \phi_1)$ needed to create a unit volume of a blend with composition ϕ from a bulk reservoir kept at composition ϕ_1. In terms of $\Delta f(\phi; \chi, \phi_1)$ the volume fraction gradient $\nabla\phi$ may be presented in the very compact form as

$$\frac{d\phi}{dz}(\phi) = \mp\sqrt{\Delta f(\phi; \chi, \phi_1)/\kappa(\phi)} \tag{8b}$$

and the total excess free energy for the equilibrium profile (interfacial tension) F_e as:

$$\frac{F_e}{k_B T A/\Omega} = \int_{\phi_1}^{\phi_2} 2\sqrt{\kappa(\phi)\Delta f(\phi; \chi, \phi_1)}\,d\phi \tag{9}$$

The upper and the lower sign in Eq. (8b) corresponds to the case when the coexisting phase ϕ_2 ($\phi_2 > \phi_1$) is located on the left-hand side and on the right-hand side of the interface, respectively.

Equation (8b) yields the "phase portrait", i.e., the $\nabla\phi$ vs ϕ dependence [61] which, integrated, results in the function $z(\phi)$ defining the equilibrium profile $\phi(z)$ (see Fig. 2). The "intrinsic width" of the interface is commonly characterized by two related quantities, $w_{1/2}$ and w, defined in terms of the extreme value of the "phase portrait" $|\nabla\phi(\phi)|_{max}$:

$$w_{1/2} = \frac{1}{2}|\nabla\phi|_{max}^{-1} \tag{10a}$$

and

$$w = \frac{\phi_2 - \phi_1}{2}|\nabla\phi|_{max}^{-1} \tag{10b}$$

While the half-width $w_{1/2}$ is related to the maximum slope, only the interfacial width w also takes into account the relevant concentration range (see Fig. 2). An analytical expression of the equilibrium profile $\phi(z)$ may be obtained only near the critical point of the symmetric binary mixture ($N_A=N_B$) with ϕ-independent interaction parameter χ [49]:

$$\phi(z) = \frac{1}{2}[(\phi_1 + \phi_2) \pm (\phi_1 - \phi_2)\tanh(z/w)] \tag{11}$$

where the characteristic width of the tanh-profile is equal to the interfacial width w from Eq. (10b) and, in this region, expressed by

$$w = \frac{\sqrt{2}}{3}a(\chi - \chi_c)^{-1/2} \tag{12}$$

The signs in Eq. (11) are governed by the same convention as for Eq. (8b). It has been also shown [49] that in the neighborhood of T_c the interfacial width w is related to the correlation length ξ_b calculated at coexistence conditions w= $2\xi_b=2\xi(\phi_1)=2\xi(\phi_2)$. In practice the hyperbolic tangent function turns out to be also a very good approximate form in the case of $N_A \neq N_B$ and ϕ-dependent χ.

2.2
Review of Experimental Results

2.2.1
Phase Coexistence in Thin Films

Until recently only *indirect* techniques have been used to determine the coexistence conditions of the high polymer mixtures: in neutron or light cloud point methods [9, 62–64] the binodal curve is approached from one-phase region at fixed overall composition of the studied blend. Crossing into the miscibility gap of the phase diagram results in the nucleation of the minority phase in the supersaturated continuum and in the observed onset of turbidity. The uncertainty of coexistence temperatures determined by this dynamic method depends on the cooling (or heating) rate and the molecular mobility of a studied mixture. A closely related method observes the onset of characteristic features in the thermogram of differential scanning calorimetry [65–67]. In another approach the volume fraction fluctuations near the critical point in the one-phase region are probed by light or neutron scattering [30, 31, 43, 50, 68–73] and the determined intensity data are analyzed in order to determine the interaction parameter χ_{SANS}. χ_{SANS} is defined within the model used for the data analysis (see Eq. 5); it is hence dependent on the model assumptions (besides commonly used incompressible blend assumption a new approach emerged relaxing the incompressibility constraints [42]) and not equivalent to the bulk interaction parameter $\chi(\phi)$.

A new approach, *directly* determining the coexistence conditions, has been proposed by us [74] and independently by Bruder and Brenn [75], providing an alternative to the various techniques mentioned above (see [19, 76, 77] for a description of works preceding the determination of this approach). The idea of this approach is to measure the profile $\phi(z)$ of two coexisting phases ϕ_1 and ϕ_2 forming a bilayer structure in a thin polymer film (see Fig. 2a). Such an equilibrium situation may in principle be obtained in the course of two processes: relaxation of the "internal" interface between pure blend components and surface induced spinodal decomposition occurring in an initially homogeneous film.

2.2.1.1
Interfacial Relaxation Leading to Coexisting Phases

Thin polymer films composed of two layers with different composition have been used for almost two decades to determine the diffusion coefficient [12, 78–82] on the basis of observed broadening of their initial profiles $\phi(z)$. When the two layers are built of two fully miscible phases (T>T_c regime for blends with upper critical solution temperature UCST), a free interdiffusion takes place with the interface growing with time t as $w_{1/2} \propto t^{1/2}$. This process proceeds without limits and results in a single homogeneous phase.

A different behavior is observed [76] for bilayers composed of partially miscible polymers below their critical temperature T_c. In this case two pure blend components interdiffuse until the equilibrium of two coexisting phases is established. The above equilibrium state is characterized by the coexistence compositions ϕ_1 and ϕ_2 and the interfacial width w. The relaxation of the initial interface between pure constituents involves two processes (see Fig. 3):
1. Formation of the equilibrium interface characteristic for coexisting phases
2. Mass transport across the interface until coexisting concentrations are reached

The first phenomenon has been analyzed theoretically by various workers [83–86]. Most of them start with the Cahn-Hilliard equation relating the flux J of the polymer species A:

$$J = -\Lambda(\phi) \frac{\partial}{\partial z} \left(\frac{\delta(F_b)}{\delta\phi(z)} \right) \tag{13}$$

with the functional differentiation of the free energy functional F_b. $\Lambda(\phi)$ is the ϕ-dependent mutual mobility [87]. The observed [88] formation of the interface, with the apparent power law dependence $w_{1/2} \propto t^\alpha$ and $0.25 < \alpha < 0.30$ (the values given here were obtained [88] by re-analyzing the data from [76, 89]), seems to match only the predictions of the theory by Harden [84]. This theory finds the bulk (predominantly entropic) terms of the free energy functional F_b to be responsible for $w_{1/2} \propto t^{1/2}$ behavior at T>T_c (in miscible regime) whereas the non-local $(\nabla\phi)^2$ gradient term of F_b is found to be in charge of the relation $w_{1/2} \propto t^{1/4}$ predicted at T<T_c (in incompatible regime) for late times of the interfacial growth.

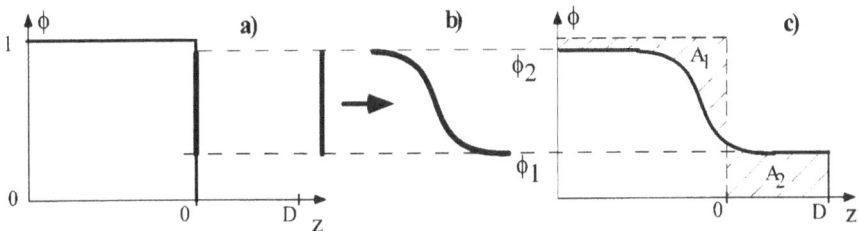

Fig. 3.a Two processes involved in the relaxation of the interface between two pure component layers (with volume fraction $\phi=1$ and $\phi=0$). **b** Interface formation. **c** Mass transport
across the interface. The interface position of the profile (*solid line* in c) shifts from the initial position (*dashed line* in c) until the *shaded areas* A_1 and A_2 match

Interdiffusion in a one-phase region is driven by bulk F_b terms. Mixing is favored by the translational entropy while it is opposed by the enthalpic interactions. These two forces may balance each other, resulting in zero net interdiffu-
sive transport at the critical point. This situation may be generalized by assuming that a diffusion coefficient goes to zero for $\phi_1<\phi<\phi_2$, between two coexisting
compositions. This is a starting point of the model proposed by Crank [90] ex-
plaining mass transport across the interface between two pure components of an
asymmetric blend with $\phi_1\neq1-\phi_2$. This model neglects the interface formation
and assumes the diffusion coefficient D_0 having a non-zero value only outside
the miscibility gap (i.e., for $0<\phi<\phi_1$ and $\phi_2<\phi<1$). It finds the unbalanced flux of
polymers interpenetrating the interface to be compensated by the interfacial
shift:

$$\frac{dI}{dt}(\phi_2 - \phi_1) = D_0 \frac{\partial \phi_{IR}(z,t)}{\partial z}\bigg|_I - D_0 \frac{\partial \phi_{IL}(z,t)}{\partial z}\bigg|_I \qquad (14)$$

where $I(t)$ denotes the position of the interface at time t while $\phi_{IL}(z,t)$ and
$\phi_{IR}(z,t)$ are the composition profiles at the left- and right-hand side of the interface $I(t)$. The solution of this model predicts the interface shift to be proportional to the square root of diffusion time t. Such a behavior has been observed experimentally [75] for finite bilayers, and also at late diffusion times where Crank
approach is not valid.

In practical terms the coexistence conditions are determined as follows. A
layer of some hundreds of nm thick of one of pure components (B say) is spin
cast on a silicon wafer with a native oxide or covered with an evaporated metal.
A similar layer of the other component (A) is laid on the top of the precast film
B using standard [74] – or modified [91] (for hydrophobic polymers) – floating
techniques. It is possible to ensure that the surface segregation and wetting effects do not perturb final phase configuration by arranging the surface preferred
component to be located near this surface. Also the substrate may be modified
(by metal evaporation) to cancel the possible polymer-substrate interactions

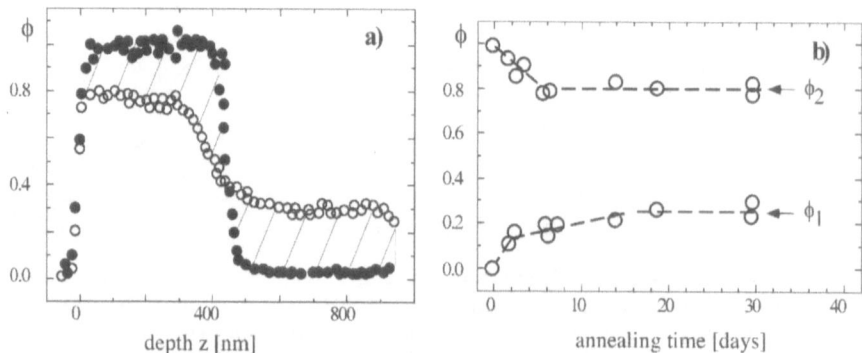

Fig. 4.a Typical concentration vs distance profiles $\phi(z)$ of the region about the interface for a hPS(m.w.=2.89×10^6)/(dPS m.w.=1.03×10^6) bilayer [74]: before annealing (●) and after 29.7 days of annealing at 170 °C (○). The absolute values of the volume fractions are determined by mounting a pure dPS layer on the top of the annealed bilayer to establish the level ϕ=1.0. **b** The variation of the plateau volume fractions ϕ as a function of annealing time [74], shown for annealing temperature 170 °C. The plateau volume fractions correspond to coexisting ϕ_1 and ϕ_2 compositions

driving the segregation to this substrate [92]. At temperatures above the glass point T_g but below T_c the two components interdiffuse to reach the equilibrium situation with formed, preferably large, composition plateaus ϕ_1 and ϕ_2 traced by depth-profiling methods.

The example of such a procedure is shown in Fig. 4a for the bilayer formed by regular (protonated) polystyrene hPS (with molecular weight m.w.=2.89×10^6) and its deuterated counterpart dPS (m.w.=1.03×10^6) [74]. The initial profile (●), corresponding to the non-annealed sample, differs considerably from that describing the situation close to the equilibrium (○) and measured on the sample annealed at 170 °C for almost 1 month. The hatched area between these two profiles marks the material transported across the interface before the coexistence situation, accompanied by the visible interfacial shift, was reached. Two types of experiments may assure us that equilibrium is reached: first, identical samples are annealed for increasing period of time and their plateau compositions are plotted vs time (see Fig. 4b) – the long time limits of the plateau compositions define the coexisting compositions ϕ_1 and ϕ_2; second, the relative thicknesses of the as-cast pure component layers in the bilayer may be varied as well as the starting compositions of the two films (within values outside the miscibility gap). As a result the amount of the material to be transported across the interface may be changed and hence the effective time needed to reach the equilibrium may also be varied.

Since the interface relaxation method was established it has been used to determine the coexistence conditions for over 20 polymer mixtures [74, 75, 88, 91, 92, 95–99].

2.2.1.2
Surface Induced Spinodal Decomposition Leading to Layered Coexisting Phases

When a binary mixture is quenched from a single homogeneous phase to a point inside the spinodal curve (see Fig. 1) a spontaneous phase separation takes place with several distinguished time regimes [49, 55, 100–103]. In the "very early stage" composition fluctuations with characteristic constant wave number q_m are magnified exponentially with time. Created disordered bicontinuous two-domain structure may be presented as a sum of composition waves with a constant wave number q_m but with *random* directions and phases. In the next, "intermediate", stage both the wavelength $2\pi/q_m$ and the amplitude of the fluctuations increase with time and the initial morphology coarsens. Finally, the compositions of created domains reach the coexisting values ϕ_1 and ϕ_2. In the subsequent processes of the "late" stage of spinodal decomposition the average domain size R(t) grows by diffusion as $R(t) \sim t^{1/3}$ and by hydrodynamic flow as $R(t) \sim t$. This description characterizes in short the spinodal decomposition occurring in the bulk.

A thin film geometry introduces two "external" interfaces, in literature called "walls", confining the polymer blend: the polymer/vacuum interface (interface denoted "I" in Fig. 5) as well as the polymer//substrate interface (denoted "II" in Fig. 5). Close to the surface the translational and rotational symmetry of the bulk fluid is broken and no concentration flux can occur through the surface. Even more important, the surface may preferentially attract one of the blend components, in consequence driving its surface segregation (described by parameters discussed in Sect. 3). This fixes the direction and the phase of the composition waves starting from the very early stage of spinodal decomposition [104–106]. As a result a coherent damped composition wave perpendicular to the surface is created as has been observed only in recent years [108–110]. This phenomenon is known as a surface-induced spinodal decomposition. Critical film thickness was observed below which the early stage wavelength $2\pi/q_m$ is changed to adapt the film size [111–113]. The separate spinodal waves, originating at two interfaces I and II, superimpose as has been concluded on the basis of their destructive interference observed for very thin films [111]. As a result a few

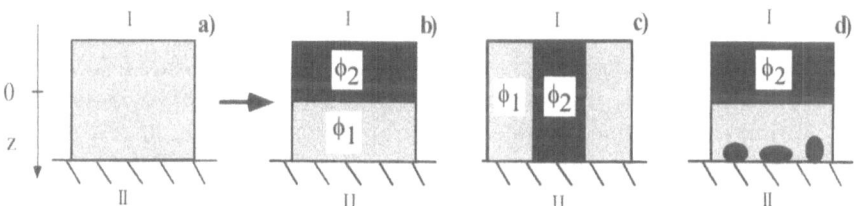

Fig. 5.a A homogeneous monolayer confined by two "external" interfaces: I and II ordering spinodal waves along z, and morphologies resulting from the spinodal decomposition. **b** Bilayer equilibrium structure [93]. **c** Equilibrium structure with two-dimensional domains [60]. **d** Exemplary transient morphology [94]

layered domains inside the film are created at later times and the domains located at both surfaces already have coexistence compositions. These two surface domains are increased at the expense of the inner ones [114] to create the equilibrium two layer coexistence morphology of Fig. 5b characteristic for strong and opposing (antisymmetric) surface fields exerted by walls I and II, each completely wetted by one of blend components [114]. An identical morphology is concluded for the pair of strong and zero surface fields [115] as it is shown later in this section.

The two scenarios described above occur together [116]: thermal fluctuations drive isotropic bulk spinodal decomposition while the surface fields favor anisotropic surface-oriented decomposition. The real situation is sometimes very complicated and it may be studied only by quasi-three-dimensional profiling [94, 114, 117] which combines the analysis of the average composition profiles $\phi(z)$ normal to the surfaces and the lateral structures yielded by microscopic methods. The depth profiling method may be used to determine unequivocally the coexistence conditions of the blend at the final stage of spinodal decomposition *only* if the bilayer morphology of Fig. 5b may be concluded. This corresponds to the absence of lateral structures on the micrographs. The lateral structures present in micrographs suggest other equilibrium [60, 118] or even transient morphologies outlined in Fig. 5.

The variation of the chemical composition of the substrate (not realized in a continuous tunable fashion) leads to drastic modifications of surface fields exerted by the polymer/substrate (i.e., II) interface [94, 97, 111, 114, 119]. The substrate may, for instance, change contact angles with the blend phase from zero to a finite value. As a result the final morphology changes from a layered structure of Fig. 5b into a column structure of Fig. 5c [94, 114]. On the other hand our very recent experiment [16] has shown that the surface fields are temperature dependent. Therefore, although it has been shown that surface-induced spinodal decomposition yields coexisting bilayer structure (Fig. 5b) at a singular temperature [114, 115], that in principle may not be necessary true for other temperatures. This motivated our comparative studies [107] on coexistence compositions determined with two techniques described above: interfacial relaxation and spinodal decomposition.

The binary mixture we used is composed of partly deuterated $E_{14}EE_{86}$ (d86) and protonated $E_{25}EE_{75}$ (h75) random olefinic copolymers characterized in [16, 91]. Here E and EE are the linear (C_4H_8) and branched ($C_2H_3(C_2H_5)$) ethylene groups, respectively. The I interface is vacuum while the II surface is always an evaporated high-purity smooth gold layer (thickness ~20 nm). Previous studies [16, 120] have determined a preferential segregation of d86 to the free surface and no detectable segregation to the Au interface. Therefore bilayers with h75 and d86 films adjacent to Au- and vacuum-interfaces, respectively, have been used in the interfacial relaxation method. The overall bilayer thickness was ~900 nm. Figure 6a presents the corresponding initial d86 profile $\phi(z)$ as a dashed step function. Annealing leads to the coexistence profile denoted by $\bigcirc$ with characteristic coexistence compositions ϕ_1 and ϕ_2. In turn, monolayers ho-

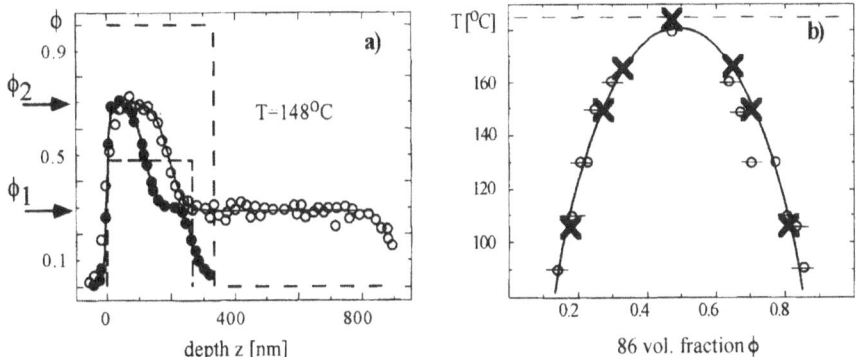

Fig. 6.a d86 volume fraction ϕ vs depth z profiles of: the d86/h75 blend monolayer with $<\phi>=0.478$ prior (- - -) and following ($\bullet$) the annealing, as well as of the d86//h75 bilayer prior (- - -) and following ($\bigcirc$) the annealing process [107]. In both cases 2 h of annealing at 148 °C was applied. Interfacial width (Eqs. 8 and 10) w (148 °C)=30 nm. **b** The coexistence concentrations determined by the relaxation of the bilayer "internal" interface [16, 91, 107] ($\bigcirc$) and by the surface induced spinodal decomposition of the monolayer [16, 107] ($\times$) yield identical binodal line (*solid curve*) described by $\chi=(0.559/T+8\times10^{-5})$ $(1-0.057\phi)$. The *dashed horizontal line* in b denotes films with homogeneous composition. Both blend components are characterized in the caption to Fig. 9

mogeneous in composition have been used in the spinodal decomposition process with corresponding initial profiles as denoted by a dashed hat function in Fig. 6a. Monolayers with constant average composition $<\phi>=0.478$ and overall thickness D, varying in the 260–600 nm range, have been obtained by spin casting a toluene-blend solution directly on the Au substrate. Subsequent annealing at temperatures below the critical point T_c has led to the coexistence profile denoted by $\bullet$ in Fig. 6a. Interference microscopy showed the lateral homogeneity of studied samples. Determined coexistence compositions ϕ_1 and ϕ_2 do not depend on the monolayer thickness in the given range. Their values are marked as $\times$ in the corresponding phase diagram in Fig. 6b. The final bilayer structure has not been observed for a monolayer annealed at $T>T_c$, i.e., corresponding to the one-phase region. The coexistence values ϕ_1 and ϕ_2 yielded by interfacial relaxation process are presented in Fig. 6b as $\bigcirc$ points. They coincide with those obtained in the course of spinodal decomposition. We may therefore conclude that both methods yield equivalent results on the coexistence concentrations.

Surface induced spinodal decomposition leads, for properly controlled surface fields, to a two layer structure characteristic for coexisting phases. Hence it may be used to determine the coexisting conditions in a more convenient way that with the interfacial relaxation method as the initial bilayer geometry may be avoided. In practical terms the overall composition of the whole thin film may be much better controlled in experiments involving spinodal decomposition. Therefore in experiments studying the equilibrium composition vs depth profiles in systems with confined geometry (i.e., in very thin layers) the spinodal decomposition is used as a route to equilibrium [121].

2.2.2
Coexistence Curve and Interfacial Width

The bilayer composed of two coexisting phases is characterized by two types of parameters. These are the compositions of coexisting phases ϕ_1 and ϕ_2 and the spatial extension of the interface separating these two phases. Experimental depth resolution ρ of an applied profiling technique is not a prime concern when measuring the profile $\phi(z)$ plateaux yielding concentrations ϕ_1 and ϕ_2. However the interfacial width w may be evaluated properly only if the resolution ρ is assured to be smaller than the measured width w_{exp}. A quadratic correction procedure is used as a standard to extract w from w_{exp}:

$$w = \sqrt{w_{exp}^2 - (\frac{6}{\pi^2 \ln2})\rho^2}$$

(15)

It is of interest to trace the temperature dependence of these two types of observables. The locus of ϕ_1 and ϕ_2 as a function of temperature yields the coexistence curve in the composition-temperature plane for the studied binary blend. A related data set on the interfacial width is plotted as a function of temperature.

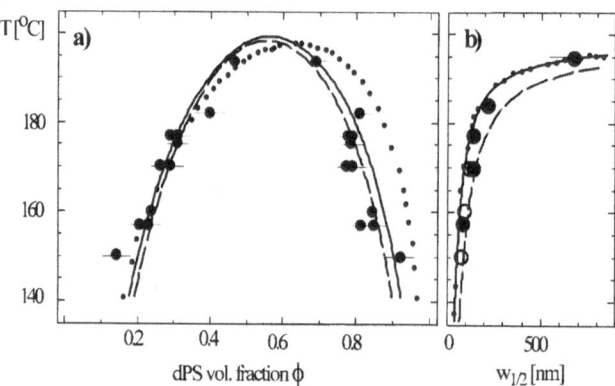

Fig. 7.a The coexistence curve. **b** The interfacial half-width $w_{1/2}$ vs temperature. Both determined for the isotopic hPS (m.w.=2.89×10^6)/dPS (m.w.=1.03×10^6) blend. Presented data show limiting values obtained in the course of the interfacial relaxation process depicted in Fig. 4: values obtained [74] by fitting a tanh to composition profiles are marked as ● whereas those [76] yielded by the steepest slopes of composition profiles are denoted as ○. The *dotted, dashed and solid curves* are generated by the following interaction parameters: $\chi = (.2/T-2.9\times10^{-4})$, $\chi=(.124/T-1.06\times10^{-4})(1-.18\phi)$, $\chi=(.18/T-2.9\times10^{-4})$ $(1+1.05\phi(1-\phi))$, respectively. Long wavelength limit $2w_{1/2}>aN^{1/2}$ holds even for longer hPS chains $(aN^{1/2}\approx112$ nm$)$

Two examples of the coexistence curve determined by us and the temperature dependence of the interfacial half width $w_{1/2}$ related to it are shown in Figs. 7 and 8. The results presented have been obtained for the isotopic mixture of protonated hPS and deuterated polystyrene dPS in the course of the interfacial relaxation process. Figure 7 corresponds to the hPS (m.w.=2.89×10^6)/dPS (m.w.=1.03×10^6) blend [74] whereas Fig. 8 to even more asymmetric hPS (m.w.=20×10^6)/dPS (m.w.=0.55×10^6) mixture [107]. All observables (ϕ_1 and ϕ_2, w_{exp}) are obtained by fitting to each measured profile a tanh-function of Eq. 11 except for the data marked by $\bigcirc$ in Fig. 7b determined numerically from the steepest slope for each profile [76]. The hyperbolic tangent describes well the experimental profiles $\phi(z)$ obtained (see Figs. 4a and 8a). It can be rigorously proven that for an even more asymmetric couple the profile $\phi(z)$ obtained numerically using Eq. (8) is well approximated by tanh-function: the concentration misfit between both curves is smaller than errors related to counting statistics of the depth profiling method. The ϕ_1, ϕ_2, and w_{exp} values taken in a long annealing time limit are assumed to describe the equilibrium situation.

Both the equilibrium interfacial width and the coexistence compositions vary with the interaction parameter χ (see Eqs. 8 and 10). The parameter χ is, in turn, temperature dependent. The coexistence conditions, such as those of Figs. 7 and 8, may be described with different level of precision by various approaches yielding different forms of the χ parameter [9].

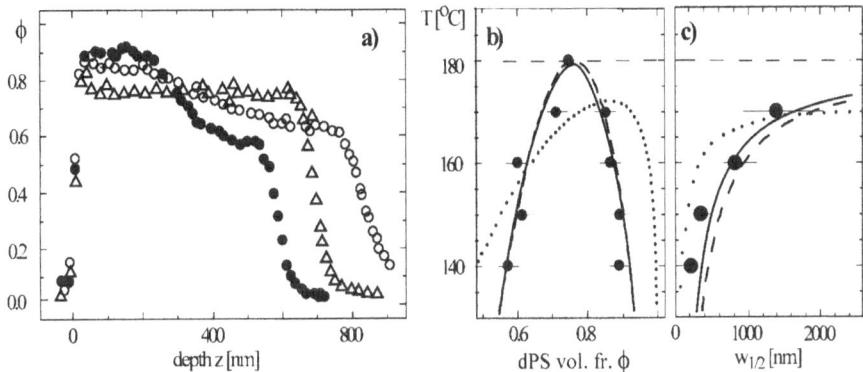

Fig. 8.a Typical composition vs depth profiles $\phi(z)$ obtained for bilayers formed by films of pure dPS (m.w.=5.5×10^5) and of 52% dPS (m.w.=5.5×10^5)/48% hPS (m.w.=20×10^6) annealed for: 28 days at 140 °C (●), 27 days at 160 °C (○) and for 14 days at 180 °C (△). **b** The coexistence curve. **c** The interfacial half-width $w_{1/2}$ vs temperature dependence, both corresponding to the interfacial relaxation process of a. The experimental coexistence conditions (●) of the asymmetric hPS (m.w.=20×10^6)/dPS (m.w.=5.5×10^5) blend described by the *dotted, dashed and solid curves* are generated by the following interaction parameters: $\chi=(.19/T-2.9\times10^{-4})$, $\chi=(.032/T+1.29\times10^{-4})(1-.22\phi)$, $\chi=(.035/T+0.65\times10^{-4})$ $(1+0.3\phi(1-\phi))$, respectively. Long wavelength limit holds even for longer hPS chains: $2w_{1/2}>aN^{1/2}\approx290$ nm. The *dashed horizontal line* in b and c denotes films with homogeneous composition

For non-polar molecules (where $\varepsilon_{AB}=(\varepsilon_{AA}\varepsilon_{BB})^{1/2}$) the simplest expression for χ, postulated by Flory and Huggins (FH) (Eq. 2), may be re-expressed as [122, 123]:

$$\chi = \frac{V}{k_B T}[\delta_A - \delta_B]^2 \tag{16a}$$

in terms of the solubility parameter δ_i [124] related to the square root of the cohesive energy density $\delta_i=[-(z_b-2)\varepsilon_{ii}/(2\,V)]^{1/2}$. Conceptually similar origin has the equation obtained by Bates et al. [69] for isotopic mixtures:

$$\chi = \frac{const_1}{k_B T}[(\frac{\alpha}{V})_A - (\frac{\alpha}{V})_B]^2 = \frac{const_2}{k_B T}[(\frac{\Delta\alpha}{\alpha})-(\frac{\Delta V}{V})]^2 \tag{16b}$$

where the contact (London dispersion) energies ε_{ij} in the original FH equation (Eq. 2) have been expressed by polarizability densities $(\alpha/V)_i$ as $\varepsilon_{ij}=const\,(\alpha/V)_i$ $(\alpha/V)_j$. The reduction in C-H bond length induced by deuterium substitution results in both the segmental volume reduction $\Delta V/V$ and the decrease of the molecular polarizability $\Delta\alpha/\alpha$ (the relative polarizability variation seems to be larger than the relative change in volume, e.g., for toluene $\Delta\alpha/\alpha\sim1.5\%$ but $\Delta V/V\sim0.3\%$). In another work Bates and Wignall [68] treat separately the volume and polarizability changes.

For isotopic blends both approaches by Bates et al. [68, 69] estimate properly the T^{-1} dependent (enthalpic) term of the more realistic interaction parameter form:

$$\chi = \frac{A}{T} + B \tag{17a}$$

commonly used by experimenters. The B term in Eq. (17a) accounts for the non-combinatorial entropy originating from nonrandom mixing due to anisotropic segment structures [10, 68]. The form of Eq. (17a) allows already for the phase diagrams with lower critical solution temperature (LCST) (for A<0 and B>0) besides those with upper T_c (UCST) considered in this work. Comparison of the experimentally determined (Figs. 7 and 8) and calculated (using Eq. 17a) coexisting conditions (dotted curves in Figs. 7 and 8) reveals two features. First, the experimentally observed behavior is reasonably predicted on a *qualitative* level. Second, it is not possible to improve the predicted fit to experimental data in any significant manner. Noticeably, the critical compositions $\phi_c=0.58(2)$ and $0.76(2)$, estimated from the data in Figs. 7a and 8b respectively, differ from the corresponding values $\phi_c=0.64$ and 0.86 obtained from the critical point condition (Eq. 3c) for a ϕ-independent interaction parameter χ.

The higher precision in describing the experimentally determined coexistence conditions is obtained by assuming that the χ parameter is ϕ-dependent. This has been known at least from the late 1970s (see [9] and references therein). Recently the interaction parameters, linear [125, 126] as well as parabolic with upward [43] and downward [70, 73] curvatures, have been reported by a number of experimental groups. Current theoretical approaches [39, 42–48, 127], which treat the effects not included in the original FH lattice model (such as segment correlations, compositional fluctuations, disparity in segmental size and shape of blend components, and their finite compressibilities), also indicate an effective interaction parameter which is ϕ-dependent. They are able to account for the observed types of $\chi(\phi)$ relation expressed by the following equations:

$$\chi = (\frac{A}{T} + B)(1 + C\phi) \tag{17b}$$

and

$$\chi = (\frac{A}{T} + B)(1 + C\phi(1 - \phi)) \tag{17c}$$

Each of the forms written above improves considerably the fit to the observed coexistence conditions of Figs. 7 and 8 (in fact $\chi(\phi, T)$ has been adjusted to fit the coexistence compositions and then the $w_{1/2}$ vs T relation was calculated). Both $\chi(\phi)$ types describe the data comparably well and therefore the linear form may be preferred as the simplest χ modification. Independent neutron scattering studies [70, 73] performed for the isotopic polystyrene mixture suggest however rather a parabolic downward $\chi(\phi)$ relation (see Sect. 2.2.4.). The magnitude of the (parabolic) χ parameter determined for both hPS/dPS blends of Figs. 7 and 8 does not differ significantly. This suggests that χ may be independent of the molecular weight [73].

Recently some of our results [74] presented in Fig. 7 were analyzed by Dudowicz et al. [47]. In their "lattice cluster model" each segment can occupy several lattice sites in order to express the segment molecular structure and local correlations. Incompressibility is lifted and unoccupied lattice sites are introduced. The related theory [128] of interfacial properties independently describes the composition profiles of both blend components. Computations [47] performed by Dudowicz well evaluate qualitatively the coexistence curve, the interfacial width as well as the corresponding ϕ-dependent effective SANS interaction parameter [73] by very similar sets of three contact (van der Waals) energies ε_{HH}, ε_{DD}, and ε_{HD}.

The mean field Cahn-Hilliard approach (Eq. 7) describes the "intrinsic" profile $\phi(z)$ about the "internal" interface between two coexisting phases. It involves only one dimension, i.e., depth z, as a lateral homogeneity is assumed [7]. Capillary wave excitations may however cause lateral fluctuations of the depth $I_e(x,y)$ at which the "internal" interface is locally positioned. As a result the effective interfacial width may be broadened beyond its "intrinsic" value (Eqs. 10 and 12). The mean field theory predicts the temperature dependence of the "intrinsic" width in a good agreement with experimental data presented here and reported by others (e.g., [76, 89] reanalyzed by [88] or [96, 129]). Some other experimental results [95, 97, 98] indicate the width larger than its "intrinsic" value (suggesting broadening due to capillary waves). This effect is more clearly visible in very recent, more precise, experiments [121, 130] tracing the dependence of the width on the film thickness (see Sect. 3.2.2).

2.2.3
Effect of Deuterium Substitution on Compatibility Conditions

In recent years many experiments, involving depth profiling techniques (ion beam methods [19, 22], SIMS [26], infrared microdensitometry [79]) and neutron scattering [131] widely use the method of "staining" individual molecules by deuterium labeling to investigate polymer chain conformations, blend miscibility conditions, polymer diffusion, macroseparation, and microsegregation. These experiments usually assume that the replacement of one of the components with its deuterium-labeled counterpart is of no significance to the system. This is not necessarily true. Already the isotopic blends, composed of deuterated and protonated polymers of otherwise identical chemical structure, exhibit [31] an upper critical solution temperature (UCST) behavior. The first directly determined phase diagrams of such isotopic blends [74], presented above, confirm the magnitude of isotopic interaction parameter $\chi_{h/d} < 10^{-3}$ yielded by SANS [43, 68–70, 73]. Deuterium labeling also affects the (bulk) compatibility conditions of mixtures composed of polymers with different chemical microstructure, as will be shown below. The influence of isotope labeling on the *surface* interactions in blends with microstructurally identical and different constituents will be discussed in Sect. 3 of this review.

Since the first report by Koningsveld et al. [132] researchers [72, 125, 133–135] have been aware that the isotope labeling effect can significantly modify the interaction parameter characterizing polymer blends. The systematic studies of this problem have been recently performed by three groups (Evanston [136], Princeton-Exxon [137–142], Weizmann Institute-Exxon [91, 96, 143]) for the binary mixtures of statistical olefinic copolymers of the structure $(E_{1-x}EE_x)_N$ where linear (C_4H_8) ethylene (E) and branched ($C_2H_3(C_2H_5)$) ethyl ethylene (EE) groups are randomly distributed in the ratio $(1-x)$: x along the chain. In these model blends of $E_{1-x}EE_x$ copolymers with different fraction x the extent of the (bulk and surface) interactions may be tuned through judicious choice of the two x values. The investigations were initiated by an interesting experimental observation made independently by all three groups [96, 136, 137], and illustrated in Fig. 9 by later [91] more spectacular result: the coexistence curve ○ determined for the hx_i/dx_j mixture of a protonated polyolefine with x_i ethyl ethylene content (we shall use the abbreviation $hx_i \equiv hE_{1-xi}EE_{xi}$ for such a fully protonated copolymer) and its partially deuterated counterpart with x_j composition ($dx_j \equiv dE_{1-xj}EE_{xj}$) is *different* from the binodal line ● concluded for the dx_i/hx_j blend, i.e., for the same couple but with reversed (swapped) isotope labeling. Here $x_i = 0.86$, $x_j = 0.75$, $N_i = 1520$, and $N_j = 1625$ and for the deuterated copolymers the fraction of $e_i = e_j = 0.4$ protons was replaced by deuterium atoms. The critical temperature T_c is higher (by 84 °C!) and hence the interaction parameter χ is

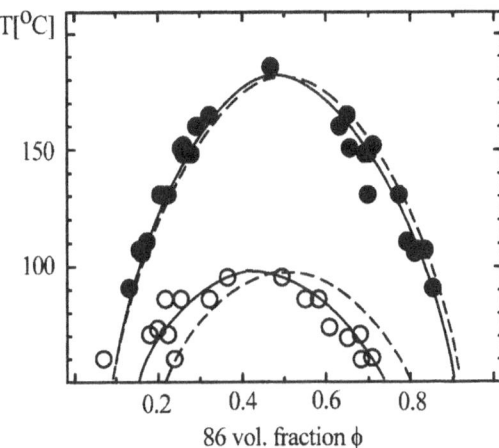

Fig. 9. Coexistence curves determined [16, 91] (see Fig. 6) for two blends (dx_i/hx_j) (●) and (hx_i/dx_j) (○). hx (dx) is shorthand for a protonated (deuterated to an extent e) copolymer $E_{1-x}EE_x$. Here $x_i=0.86$, $x_j=0.75$, $e_i=e_j=0.4$, $N_i=1520$ and $N_j=1625$. The binodals denoted by *solid lines* are generated by $\chi(d86/h75)=(.559/T+8\times10^{-5})$ $(1-0.057\phi)$ and $\chi(h86/d75)=(.547/T-8\times10^{-5})(1-0.217\phi)$ whereas those marked as *dashed lines* correspond to $\chi(d86/h75)=0.578/T$ and $\chi(h86/d75)=0.471/T$

larger for the component with larger EE fraction labeled by deuterium; i.e., $\chi(dx_i/hx_j)>\chi(hx_i/dx_j)$ for $x_i>x_j$. The observed shift in the effective interaction parameter χ by ca. 2.8×10^{-4} (at 100 °C) cannot be explained by the isotopic effect alone even if a small difference in the deuteration level is allowed (only inadmissably large value $(e_i-e_j)>0.53$ can reproduce the shift in χ).

The explanation of this puzzling phenomenon has been given by us [143] and by two other groups [136, 137] independently. All these explanations are parallel and similar in spirit. They differ rather in the number of free parameters than in mechanism claimed to be in charge of the observed effect. All approaches start from the microscopic model of Bates et al. [68, 69] relating, for isotopic mixtures, the interaction parameter χ with the slight differences in molar polarizability $\Delta\alpha/\alpha$ and segmental volume $\Delta V/V$ between two blend components. The decrease upon deuteration in C-H bond lengths, resulting in both $\Delta\alpha/\alpha$ and $\Delta V/V$, may be interpreted in terms of the solubility parameter δ being reduced (see Eq. 16 and the inset to Fig. 10a).

In the next step the variation of the solubility parameter δ is considered due to the change in the microstructure. All three descriptions agree that the parameter δ of the random copolymer $E_{1-x}EE_x$ should decrease monotonically with increasing ethyl ethylene fraction x (see the inset to Fig. 10a). The original Bates formulation are extended beyond isotopic mixture by [136, 143] (but still for nonpolar substance and similar volumes of interacting species $(V_E-V_{EE})/V\approx1.4\%<<1$) emphasizing the role of $\Delta V/V$ alone [136] or correlated $\Delta\alpha/\alpha$

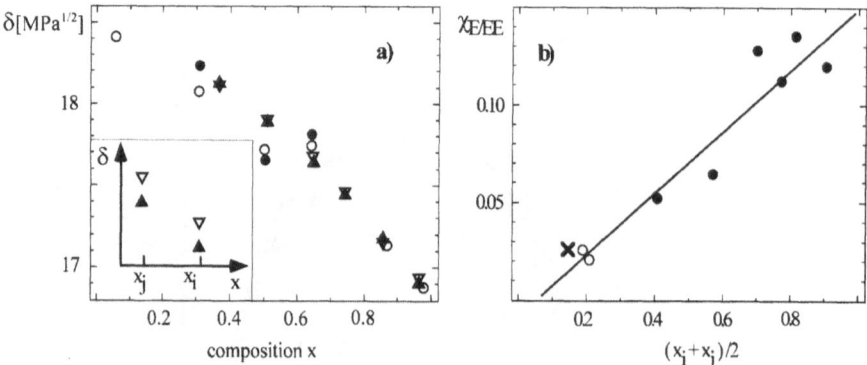

Fig. 10.a The inset shows the postulated variation of the solubility parameter δ caused by deuterium labeling (symbols ▲ and ▽ correspond to labeled and nonlabeled copolymers, respectively) and due to the change in ethyl ethylene fraction x. The cumulative analysis, described in text, yields the absolute δ value for deuterated dx (▲) and protonated hx (▽) copolymers as a function of x at a reference temperature T_{ref}=100 °C: determined interaction parameters (as in Fig. 9) allow us to determine two sets of differences Δδ adjusted here to fit independent PVT data [140, 141] measured at 83 °C (●) and at 121 °C (○). **b** The interaction parameter, $\chi_{E/EE}$, arising from the microstructural difference contribution to the overall effective interaction parameter $\chi(hx_j/dx_i,e_i)$ in Eq. (19) as a function of the average blend composition $(x_i+x_j)/2$ at a reference temperature of 100 °C. $\chi_{E/EE}$ values are calculated (see text) from coexistence data (● points correspond to [91, 143] and ○ symbols to [136]) for blend pairs, structurally identical but with swapped labeled component.✕ marks $\chi_{E/EE}$ yielded directly [134] for a blend with both components protonated. *Solid line* is the best fit to data

and ΔV/V [143] in the effective interaction parameter. Another approach [137] finds such a monotonic change in δ parameter with composition to be in accord with a group contribution method for estimating δ [124]. The tendency of the solubility parameter δ to decrease with increased deuterium extent e *or* EE fraction x (see the inset to Fig. 10a) allows us already to explain qualitatively the isotope swapping effect presented in Fig. 9:

$$\chi(dx_i/hx_j)=\frac{V}{k_BT}(\delta_{ei,xi}-\delta_{0,xj})^2 > \chi(hx_i/dx_j)=\frac{V}{k_BT}(\delta_{0,xi}-\delta_{ej,xj})^2 \quad <18a>$$

for $x_i>x_j$ as the difference $|\delta_{0,xj}-\delta_{ei,xi}|>|\delta_{ej,xj}-\delta_{0,xi}|$.

The change of the solubility parameter $\delta_{ei,xi}$ may be assumed to be linear with deuterium fraction e_i and composition x_i ($0\le e_i\le1, 0\le x_i\le1$):

$$\delta_{ei,xi}=\delta_0-x_i(\delta_0-\delta_1)-e_i(\delta_h-\delta_d) \quad (18b)$$

where $(\delta_h-\delta_d)$ is the solubility parameter difference between any protonated random copolymer $E_{1-x}EE_x$ and its deuterated counterpart. Equation (18b) only holds locally, for any specific value of x, δ_0, and δ_1 are not the values of solubility parameter for pure (protonated) polyethylene and poly(ethyl ethylene) but $(\delta_0-\delta_1)$ rather describes the δ(x) slope (see Fig. 10a).

Finally the effective interaction parameter $\chi(hx_j/dx_i,e_i)$ in a mixture of a protonated random copolymer hx_j and a partly deuterated (to extent e_i) random copolymer dx_i is written using Eqs. (16a) and (18b) for x_j value close to x_i one:

$$\chi(hx_j/dx_i,e_i) = \frac{V}{k_BT}(\delta_{0,xj} - \delta_{ei,xi})^2 = \frac{V}{k_BT}[(x_i - x_j)(\delta_0 - \delta_1) + e_i(\delta_h - \delta_d)]^2$$

$$= (x_i - x_j)^2 \, \chi_{E/EE} + e_i^2 \, \chi_{h/d} + 2e_i(x_i - x_j)\sqrt{\chi_{E/EE}\chi_{h/d}}$$

(19)

Equation (19) is the central result of our approach [143], which is the simplest theory accounting for the effect of deuterium substitution. The first two terms on the right-hand side of Eq. (19) represent the straightforward addition of the effects related to pure microstructural $\chi_{E/EE}$ and isotopic $\chi_{h/d}$ interaction parameters while the third term accounts for their "interference". This last term can enhance or diminish the effective interaction parameter $\chi(hx_j/dx_i,e_i)$ depending on the sign of the difference in composition (x_i-x_j). While the range of the microstructural interaction parameter $\chi_{E/EE} > 2\times10^{-2}$ is much higher than that of the isotopic parameter $\chi_{h/d}$ their geometrical average spans the gap between both ranges. In the other approaches [136, 137] the magnitude of the "interference" term for $x_i>x_j$ is different from that $x_i<x_j$ as more adjustable parameters are considered. Formulations similar to Eq. (19) may be obtained [136] by extending random copolymer theory [144].

The effective interaction parameter $\chi(hx_j/dx_i,e_i)$ in Eq. (19) is a function of two unknown parameters, $\chi_{E/EE}$ and $\chi_{h/d}$, so these can be extracted [143] from two experimentally determined values of the overall interaction parameter for each pair of structurally identical mixtures with a swapped isotope labeled component, i.e., for dx_i/hx_j and hx_i/dx_j. In order to fit to the specific form AT^{-1} of the interaction parameter used in the approach described above we re-express all effective interaction parameters as [91] $\chi(hx_j/dx_i,e_i)=\chi_cT_c/T$, where $\chi_c=(N_i^{1/2}+N_j^{1/2})^2/(2N_iN_j)$ and T_c is given by experiment. It turns out [145] that this is a good approximation for studied random polyolefins (see Fig. 9) as the entropic term B of Eq. (17a,b) is small (it contributes less than 6% for most of the blends) and the weak ϕ-dependence of χ may be neglected as it is not directly involved here.

The procedure extracting $\chi_{E/EE}$ and $\chi_{h/d}$ values has been applied to six pairs of olefinic blends with coexistence conditions determined by profiling method [91, 143] and to two other pairs with T_c determined by morphology studies using scanning electron microscopy [136]. The $\chi_{E/EE}$ values related to these two data sets are marked as ● and ○, respectively, in Fig. 10b. They are plotted as a function of the average composition $(x_i+x_j)/2$ for each pair of blends. The corresponding $\chi_{h/d}$ values (not presented here) are much scattered and comparable to or even smaller than 10^{-4}, while 4×10^{-4} is obtained when analyzing the full $\chi(\phi)$ form [143]. These $\chi_{h/d}$ values may be compared with the effective interaction parameter evaluated directly by SANS [72, 136] for mixtures of protonated and partly deuterated copolymers dx_i/hx_i. SANS yields the $\chi_{h/d}$ absolute magnitude to be scattered in the range (-10^{-3}) to $(+10^{-3})$. Thus, while the calculated values of the "bare" isotope interaction are within this range a more detailed comparison is precluded [143].

The calculated values of "pure" microstructure interaction parameter $\chi_{E/EE}$, plotted as an average blend composition $(x_i+x_j)/2$ in Fig. 10b ($\bullet$ and $\bigcirc$ points), are in accord with the $\chi_{E/EE}$ value (marked as $\times$) evaluated directly [134] for a blend with both components protonated and with $(x_i+x_j)/2=0.15$. The monotonic increase of $\chi_{E/EE}$ with blend composition has been noted not only by profiling studies [91] but also earlier by SANS [139, 140]. Theoretical explanations of the $\chi_{E/EE}$ vs $(x_i+x_j)/2$ relation have been suggested recently [140, 146]. High order random copolymer theory [147] assumes that the primary interacting elements in the lattice model are not segments but rather segment pairs called diads. The diad $\chi_{E/EE}$ expression for a blend of random copolymers indeed allows for the monotonic compositional dependence. Another approach [146] suggests that the origin of the effect is in conformational and architectural "mismatch" between blend components.

The isotope swapping effect may be also used to infer the order of solubility parameters [139] for protonated blend components δ_{xi} and δ_{xj} since their interaction parameter $\chi(hx_i/hx_j)$ (see Eq. 16a) yields only the magnitude of the difference $|\delta_{xi}-\delta_{xj}|$. We have:

$$\delta_{xi}<\delta_{xj} \text{ if } \chi(hx_i/dx_j,e_j)<\chi(hx_j/dx_i,e_i). \tag{20}$$

Using this approach we determined [145], independently from assumptions made before Eq. (18b) was used, the relations between the absolute values of δ for pure random copolymers grouped in two sequences: h38-d52-h66-d75-h86-d97 and d38-h52-d66-h75-d86-h97. We adjusted them [145] to fit the best δ values yielded by PVT properties [140, 141] measured at 83 °C ($\bullet$) and at 121 °C ($\bigcirc$). The concluded absolute values of δ are presented in Fig. 10a as a function of the composition x for labeled ($\blacktriangle$) and nonlabeled ($\triangledown$) copolymers (at T_{ref} = 100 °C). We see immediately the assumed compositional variation of the solubility parameter. In addition we notice that the absolute value of the $\delta(x)$ local slope increases with composition x. This is directly related to the increase with x in "pure" microstructural interaction parameter $\chi_{E/EE}$.

2.2.4
Coexistence Conditions in Thin Films vs Those in the Bulk (Analysis with Neglected Finite Size Effects)

Equilibrium structure of two coexisting phases can be described for thin films by two types of morphologies depicted by Fig. 5b,c. They could appear in regimes defined by (see Sect. 3.2): (i) the character of both "external" interfaces (I and II in Fig. 5) exerting specific surface fields; (ii) overall film thickness D; (iii) temperature. Direct determination of coexistence concentrations in films with a lateral column-like structure of Fig. 5c would be possible only with the (expected) advent of real three-dimensional profiling techniques. So far we are able to measure directly binodal compositions only by profiling films with a bilayer morphology of Fig. 5b (see Sect. 2.2.1).

The experimental approach examines bilayers with a limited precision in depth z ($\delta \approx$ a few nanometers) and in volume fraction ϕ (a few percent). It assumes that at least the central part of the analyzed profile $\phi(z)$ describes only the "internal" interface between coexisting phases ϕ_1 and ϕ_2. This is not necessarily true when surface segregation regions, adjacent to both "external" interfaces, cannot be neglected as it is for very thin films. Related finite size effects are discussed in detail in Sect. 3.2: theoretical models and computer simulations expect that size effects modify the "intrinsic" profile $\phi(z)$ in films with thickness D not much larger than the width w of the "internal" interface. Therefore size effects may lead in principle to systematic errors [6] of binodal values determined for films which are very thin or are profiled at $T \rightarrow T_c$ (where the ratio D/w is also small due to the diverging w).

The mean field theory describes coexistence conditions with the exception of an Ising-type region located around the critical temperature T_c. The related non-mean field behavior has been observed in scattering studies [41, 148, 149]: Narrow temperature ranges of the Ising regime (e.g., 1.6° for N=4200) and even smaller shifts of T_c values from their mean field predictions (e.g., 0.2° for N= 4200) were reported for high polymers [148]. More recent analysis indicated non-mean field regions a few times larger than described above [149]. They are located in a critical region in which binodal determination in thin films is already problematic due to size effects. In fact, non-mean field behavior leads to novel effects [150] expected for confined geometry (see Sect. 3.2.2).

In this section we address the question of accordance between coexistence conditions determined for polymer mixtures in the bulk and confined in thin bilayer films. Macroscopic samples with the size of ca. 1 mm are analyzed by Small Angle Neutron Scattering. It probes the compositional fluctuations *away from binodal* to yield the effective interaction parameter $\chi_{SANS}(\phi)$. Thin bilayer films are studied by profiling techniques to yield concentrations ϕ_1 and ϕ_2 *at binodal*. These are described by the composition dependent interaction parameter $\chi(\phi)$. In fact only the section of the relation $\chi(\phi)$ bounded by ϕ_1 and ϕ_2 is relevant as it describes the whole "intrinsic" profile of Fig. 2.

Two problems have to be solved to compare the results obtained for bulky samples and for thin films.

First, different data sets (away from and at binodal) must be scaled to the identical temperature and concentrations. This is possible for a few blends with (assumed) identical interaction parameter but characterized by different chain lengths (N_A, N_B) and hence different phase diagrams. This method is used for isotopic polystyrene mixtures. If the parameter $\chi_{SANS}(\phi)$ is linear with 1/T for each concentration ϕ then $\chi_{SANS}(\phi = const, T)$ can be reasonably extrapolated to regions at or inside coexistence curve. We use this solution for olefinic blends composed of random copolymers $E_{1-x}EE_x$. Here the self-same mixtures are used in both bulk SANS samples and in profiled thin films.

Second, $\chi_{SANS}(\phi)$ is not equivalent to the interaction parameter $\chi(\phi)$. However for each type of the $\chi(\phi)$ relation (Eq. 17b,c) the corresponding effective $\chi_{SANS}(\phi)$ parameter is easily calculated (see Eq. 5):

$$\chi_{SANS}[\chi=(A/T+B)(1+C\phi)]=(A/T+B)(1-C+3C\phi) \qquad (21a)$$

and

$$\chi_{SANS}[\chi=(A/T+B)(1+C\phi(1-\phi))]=(A/T+B)(1-C+6C\phi(-\phi)) \qquad (21b)$$

allowing us to compare χ_{SANS} values corresponding to the bulk with those corresponding to thin film geometry.

The isotopic mixture of polystyrene is the first system, where both SANS [70, 73, 151] and depth profiling studies [74, 107] (Figs. 7 and 8) of coexistence conditions have been reported. The initial SANS measurements by Bates and Wignall [151] focused on the temperature dependence only, and have been repeated recently for different compositions [73] motivated by the reports of the negative effective interaction parameter being observed [70]. The results of these repeated measurements [73] performed at T=160 °C, marked by circles ($\bigcirc$ and $\bullet$) in Fig. 11, agree with the initial data [151] and show no signs of negative χ_{SANS}, very unlike for isotopic mixtures with repulsive interactions [73]. Further, they suggest no m.w. dependence as two data ($\bigcirc$ and $\bullet$) sets, obtained for two different couples, agree with each other. Evaluated by profiling techniques $\chi_{SANS}(\phi)$ values are presented in Fig. 11a,b for postulated linear and parabolic $\chi(\phi)$ relations, respectively. Solid and dashed lines correspond to two different isotopic

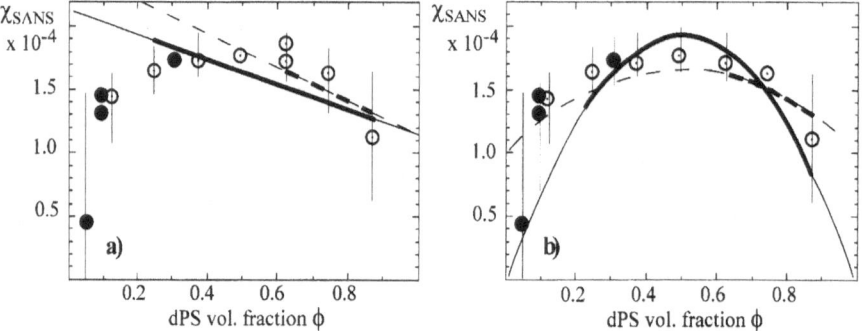

Fig. 11a,b. The relation χ_{SANS} vs dPS volume fraction ϕ determined for various isotopic polystyrene blends at T=160 °C by SANS [73] ($\bigcirc$ and $\bullet$) and evaluated [74, 107] with the profiling technique (Figs. 7 and 8) from $\chi(\phi)$ fitting the coexistence curves at a wide temperature- and volume fraction-range (*solid and dashed lines*): **a** the linear form; **b** the parabolic form, both postulated for $\chi(\phi)$. $\bullet$ and $\bigcirc$ points mark the hPS (m.w.=9.05×10^5)/dPS (m.w.=1.29×10^6) and the hPS (m.w.=1.6×10^6)/dPS (m.w.=1.29×10^6) blend data, respectively. *Solid lines* correspond to the hPS (m.w.=2.89×10^6)/dPS (m.w.=1.03×10^6) blend whereas the *dashed lines* corrspond to the hPS (m.w.=20×10^6)/dPS (m.w.=5.5×10^5) mixture. *Thin lines* denote the overall evaluated $\chi_{SANS}(\phi)$ variation while their *thickened sections* are bounded by coexisting values ϕ_1 and ϕ_2

mixtures, characterized by coexistence curves of Figs. 7 and 8, determined for bilayers with D/w ranges of 7–28 and 4–8, respectively. It is evident that both χ forms yield variations that match independent SANS data within their error bars and for the volume fractions bounded by coexistence compositions, marked by thickened sections in Fig. 11. The marked large SANS systematic error bars are typical for blends with weak interaction parameter.

The binary mixtures of protonated (hx_1) and partially deuterated (dx_2) statistical copolymers $E_{1-x}EE_x$ ($x=x_1, x_2$) of ethyl ethylene (EE) and ethylene (E) segments form the second group of systems which coexistence conditions have been studied with bulk SANS method [138] and with depth profiling techniques [91, 96] adequate for thin films. Here we compare results obtained for two blends: h88 ($N_{88}=1610$)/d78 ($N_{78}=1290$ and deuteration fraction e=0.296) and h66 ($N_{66}=2030$)/d52 ($N_{52}=1510$ and e=0.34) presented in 12a,b, respectively. Again χ_{SANS} data (marked in Fig. 12 as open symbols for different temperatures) are determined by SANS directly at each singular composition ϕ in one-phase region, i.e., outside the coexistence curve in temperature-composition plane. Additionally, χ_{SANS} values corresponding to concentrations inside the miscibility gap are extrapolated [138] from higher temperatures (and marked as filled symbols in Fig. 12). The $\chi_{SANS}(\phi)$ relation determined by SANS seems to be linear above T_c and slightly parabolic (with upward curvature) below T_c.

A depth profiling method has been used to determine coexistence curves for a large group of systems [91, 96] structurally similar to the h88/d78 and h66/d52 blends. Performed data analysis of profiled bilayers (with total D/w range 4–57) suggests [91] that while the linear $\chi(\phi)$ form can describe all experimental bin-

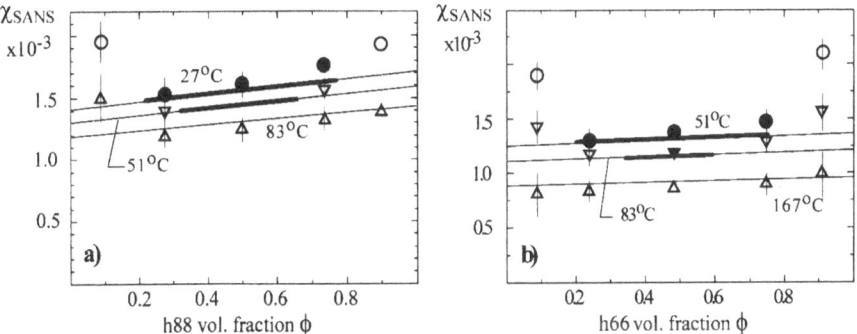

Fig. 12a,b. The relation χ_{SANS} vs the volume fraction ϕ of protonated component determined by SANS [138]: directly (*open symbols*) or extrapolated [138] at each ϕ from higher temperatures (*filled symbols*) and evaluated with the profiling technique (at three temperatures about T_c) from $\chi(\phi)$ fitting the coexistence curves [91, 96] (*solid lines*). The thickened sections of solid lines are bounded by coexisting values ϕ_1 and ϕ_2. **a** Corresponds to the h88 ($N_{88}=1610$)/d78 ($N_{78}=1290$, e=0.296) blend at 27 (O, ●), 51 (∇) and 83 °C (Δ). **b** Corresponds to the h66 ($N_{66}=2030$)/d52 ($N_{52}=1510$, e=0.34) blend at 51 (O, ●), 83 (∇, ▼) and 167 °C (Δ)

odals, this is not true for χ parabolic in ϕ. Therefore the linear $\chi(\phi)$ is used to describe coexistence curves of all olefinic blends [91, 96] including the mixtures h88/d78 and h66/d52: the solid lines in Fig. 12 mark the $\chi_{SANS}(\phi)$ relation evaluated from linear $\chi(\phi)$ for different temperatures. Their thickened sections, bounded by coexisting values ϕ_1 and ϕ_2, are in good agreement with the χ_{SANS} values extrapolated from high temperature SANS data. This accordance with SANS data is widened to almost whole concentration range at temperatures above T_c.

Similar agreement between $\chi_{SANS}(\phi)$ yielded by SANS and extracted from determined coexistence curve may also be concluded for the d66/h52 blend. We are not aware of any other polymer blends where available data would allow for similar analysis. The comparison of coexistence data characteristic for bulk and for thin film geometry discussed above, although limited only to four different mixtures, suggests their equivalence at least for concentrations bounded by coexisting compositions and within error bars, relatively large for blends with weak interaction parameter. This conclusion enhances the potential role of thin film methods, such as interfacial relaxation or surface induced spinodal decomposition, in evaluating the coexistence conditions characterizing polymer blends also in the bulk. A wide temperature range of data (bilayers with large D/w ratio mainly) used to evaluate χ and a limited precision of experimental techniques that were used precludes here an observation of size effects.

2.3
Summary and Conclusions

Coexistence conditions of high polymer mixtures may be determined directly with the advent of the novel approach [74, 75] focused on two coexisting phases confined in a thin film geometry and forming a bilayer morphology. Such equilibrium situation is obtained in the course of relaxation of an interface between pure blend components *or* in late stages of surface induced spinodal decomposition. It is shown that both methods lead to equivalent results [107] (Sect. 2.2.1).

Depth profiling techniques applied to thermodynamically equilibrated thin films characterize the compositions of coexisting phases and the spatial extent of the separating interface. This procedure repeated at different temperatures yields the coexistence curve and the corresponding temperature variation of the interfacial width. Determined coexistence curves are well described by the mean field theory with composition-dependent bulk interaction parameter [74]. The same interaction parameter also seems to generate the interfacial widths in accordance with results presented here [107] (Sect. 2.2.2) and elsewhere [88, 96, 129]. These predictions may however need to be aided by capillary wave contributions to fit another observations [95, 97, 98], especially those tracing the change of the interfacial width with film thickness [121, 130] (see Sect. 3.2.2).

The question of the accordance between the coexistence conditions determined by the novel direct approach profiling thin films and those evaluated indirectly by commonly used neutron scattering in bulk samples (SANS), is addressed in Sect. 2.2.4 [107]. The comparison between these two methods is possible only for the effective bulk interaction parameter χ_{SANS}: two corresponding sets of $\chi_{SANS}(\phi)$ values match each other (Figs. 11 and 12) within relevant error bars and for concentrations bounded by binodal values ϕ_1 and ϕ_2; this matching is more apparent for two olefinic blends (Fig. 12) where the self-same mixtures were used both in bulk SANS samples and in profiled thin films. The comparison presented enhances the potential role of the novel approach [74, 75] suggesting that it may be used to evaluate the coexistence conditions of polymer blends in the bulk.

"Staining" of individual molecules by deuterium labeling is commonly used in condensed matter studies. This technique may lead to serious consequences for macromolecular systems. It is because the threshold bulk interaction, necessary for phase separation, is easily reduced (for a large number of segments per chain) below the magnitude characteristic for weak disperse interactions between molecules and their deuterated counterparts. First coexistence curves determined directly for mixtures composed of polymers with different isotopic status are presented in Sect. 2.2.2 [74, 107]. Perturbation in compatibility conditions caused by isotope "staining" in mixtures of polymers with slightly different microstructure is, in turn, discussed in Sect. 2.2.3 [136, 137, 143]. The simple approach presented allows us to extract "pure" isotopic and microstructural contributions to the overall bulk interaction parameter for nonpolar mixtures. It explains also why the coexistence curve for a polymer mixture may be different from that obtained for the same blend but with a swapped isotope "stained" component.

3
"External" Interfaces of Binary Polymer Blends

3.1
Semi-Infinite Mixture Approach

A two component liquid mixture (A and B, say) with an upper critical solution temperature (UCST) T_c will separate, below T_c, into two coexisting phases: ϕ_1 rich in B and ϕ_2 rich in A component (see Fig. 1). Let us consider the two coexisting phases in contact with a free surface or a solid. The geometrical arrangement of these two phases in the vicinity of the surface is described in terms of the contact angle θ (see Fig. 13). When the contact angle θ is equal to zero, a layer of the phase ϕ_2 is located on the surface and completely separates the second phase ϕ_1 from it; this is designated as *complete* or *perfect wetting*. *Partial wetting* occurs for non-zero values of the contact angle.

The relation for the equilibrium contact angle was given almost two centuries ago by Young [152]:

$$\gamma_{12}\cos\theta=\gamma_{1s}-\gamma_{2s} \tag{22}$$

Here subscripts s as well as 1 and 2 refer to the surface and the phases ϕ_1 and ϕ_2, respectively, and γ_{ij} refers to interfacial (surface) tension between the i-th and

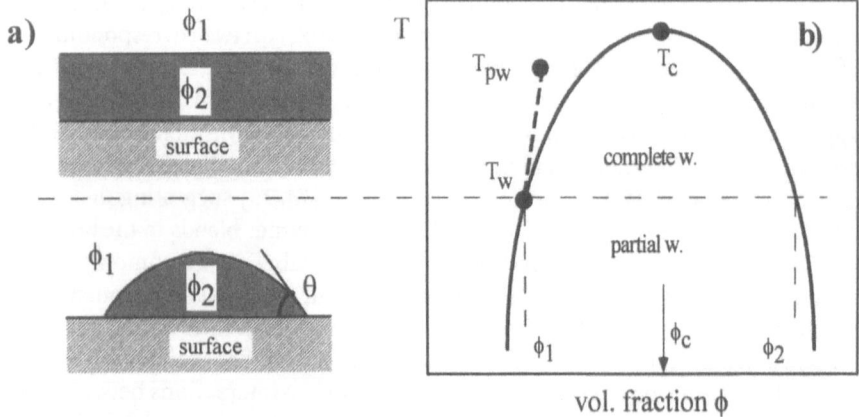

Fig. 13a,b. A binary mixture bounded by a surface: **a** two possible arrangements of coexisting phases (ϕ_1, ϕ_2) at the surface; **b** the corresponding surface phase diagram. With phase ϕ_1
in the bulk of the mixture, phase ϕ_2 wets the surface partially or completely (i.e., the contact
angle $\theta \neq 0$ or $\theta = 0$) at temperatures below or above the wetting transition point T_w. For a first
order wetting transition a prewetting line (*dashed line*), terminated by the prewetting critical point T_{pw} is present in the one-phase region

j-th phase. It is conceivable that the balance of the surface tensions, expressed by
Eq. (22), may be changed with temperature, leading to the transition from partial to complete wetting. This transition is said to occur at the *wetting temperature* T_w (see Fig. 13). If the derivative of the contact angle with temperature is
discontinuous at T_w then the transition is designated as the *first order* one. A second order transition corresponds to a continuous derivative and is called *critical
wetting*.

Physics of wetting transitions has reemerged in the focus of interest following
the seminal work by Cahn [153]. He pointed out that, while approaching the crit-
ical point T_c, the interfacial tension $\gamma_{12} \propto (T_c-T)^{2\nu}$ goes to zero more quickly than
the surface tension difference $(\gamma_{1s}-\gamma_{2s}) \propto (T_c-T)^\varepsilon$ (since the bulk critical exponent
$2\nu \approx 1.3$ is larger than the surface critical exponent $\varepsilon \approx 0.8$). Far below T_c the two
phases are incompatible so that $\gamma_{12} > (\gamma_{1s}-\gamma_{2s})$ and partial wetting is expected.
However, due to the scaling argument given above, we should have always $\gamma_{12} =
(\gamma_{1s}-\gamma_{2s})$, i.e., the complete wetting behavior in the region close enough to T_c.
This phenomenon is called a *critical point wetting*. First order wetting transitions have been determined in a variety of small molecule liquid mixtures [154–
156] but critical wetting has been observed only very recently [157].

Another insight into the wetting phenomena, alternative to that yielded by
the contact angle θ, has been given by the profile $\phi(z)$ (composition ϕ vs depth
z) [8, 53, 61, 153, 158]. The surface of a two component liquid mixture which favors one of the components will be *enriched* in that component, A say. When the
region far from the surface (bulk region) is occupied by the B-rich phase ϕ_1, the
surface concentration ϕ_s (ϕ_{1s} or ϕ_{2s} in Fig. 14) is higher than this binodal value

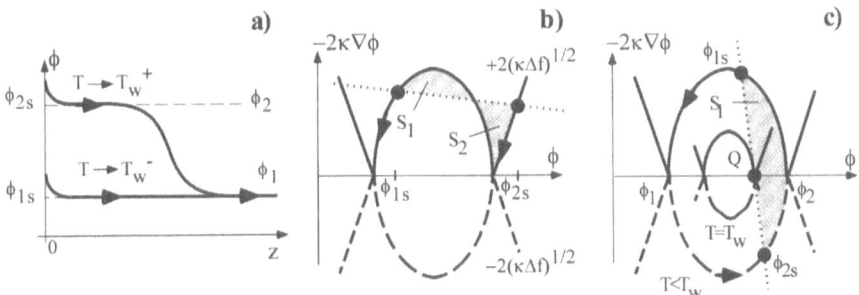

Fig. 14.a Composition-depth $\phi(z)$ profiles near the surface (at z=0) of a binary mixture at bulk concentration ϕ_1 at lower $(T{\rightarrow}T_w^-)$ and upper $(T{\rightarrow}T_w^+)$ limit of the first order wetting transition point T_w. **b,c** *Cahn constructions* with trajectories $-2\kappa\nabla\phi$ plotted for profiles $\phi(z)$ with decreasing (*solid lines*) and increasing (*dashed lines*) slopes. Surface boundary condition (Eq. 26) is met at points (marked by ●) where surface energy derivative $(-df_s/d\phi)_s$ (*dotted line*) intersects trajectories $-2\kappa\nabla\phi$ at concentrations reached at the surface. Cahn plot b corresponding to the first order transition depicted in a; Cahn construction c typical for a critical wetting: trajectories $-2\kappa\nabla\phi$ with larger extrema correspond to temperatures $T<T_w$ while those with smaller humps are for $T=T_w$

ϕ_1. Thus an excess layer exists at the "external" interface (surface) of the phase ϕ_1. For the case of partial wetting $\phi_1<\phi_s<\phi_2$ and the surface excess layer is finite. For complete wetting $1>\phi_s\geq\phi_2$ and the bulk phase ϕ_1 is completely separated from the surface by a macroscopic layer of the phase ϕ_2. The thickness of the excess layer increases to infinity as temperature approaches T_w: it diverges smoothly for a critical wetting and jumps from a finite to macroscopic value for first order wetting transition [158].

The original Cahn argument for the existence of wetting transitions in simple liquids also applies to polymer mixtures but there are two new features in the polymer case. First, there is a drastic reduction in the interfacial tension γ_{12} describing coexisting phases of binary polymer blend as compared to the mixture of chemically identical simple liquids (monomers or oligomers). This is due to polymer chain connectivity and resulting, extremely small, interaction parameter χ ($\chi_c=2/N$) and hence also small interfacial tension (see Eq. 9). On the other hand, segment-surface interactions and hence also surface tensions γ_{1s} and γ_{2s} remain comparable for polymers and for small molecules built of identical segmental units. As a consequence, the equality $\gamma_{12}=(\gamma_{1s}-\gamma_{2s})$ and the wetting transition should readily occur a far away from criticality at $T_w<<T_c$ [15]. The second new feature is that the location of the binodal line, determining the two phases ϕ_1 and ϕ_2, may be varied by the molecular weights of the polymers. Thus in principle we may have wetting transition as a function of the molecular weight [159]. No wetting transition (depicted in Fig. 13) has been reported so far for polymer mixtures, although its indirect indications have been observed by us recently (see Sect. 3.1.2.3). Complete wetting behavior has been determined in polymer

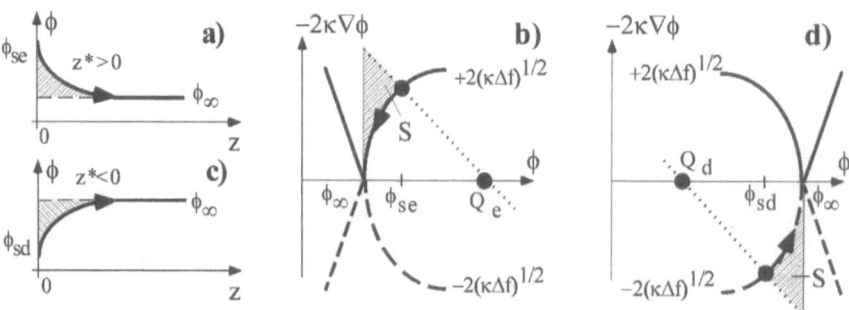

Fig. 15a–d. Composition-depth $\phi(z)$ profiles of a binary mixture with a bulk concentration ϕ_∞: **a** *enriched*; **c** *depleted* at the surface (z=0); **b,d** corresponding Cahn constructions. In a and c *hatched areas* mark the surface excess z*. z*>0 (with surface concentration $\phi_{se}> \phi_\infty$) for surface enrichment whereas z*<0 (and surface concentration $\phi_{sd}<\phi_\infty$) for surface deple-tion. In b and d *shaded areas* mark the gain in the excess free energy F_e caused by compo-sition at the surface being modified as compared to that in the bulk

mixtures [14, 16, 114] even at temperatures lower than 100° below T_c [16, 160] indicating the possibility of a drastic extension of the critical point wetting re-gime. Theoretical works yield inconsistent predictions regarding the character of the surface phase diagram for polymer mixtures [7]: for the region far below T_c first order wetting is shown to be possible by mean field theory [15] and crit-ical wetting is advocated to occur there by Monte Carlo simulation [161] while both types of wetting transition are allowed by a self-consistent mean field ap-proach [162] depending on details of surface interactions. Very recently a wet-ting *reversal* transition, rather than a wetting transition, has been reported [163] in thin polymer films: a change from a trilayer ($\phi_2/\phi_1/\phi_2$) to a bilayer (ϕ_2/ϕ_1) film morphology is concluded as temperature increases.

The blend composition at the surface ϕ_s differs from the bulk concentration ϕ_∞. Similar to the wetting layer (for $\phi_\infty=\phi_1$), a surface excess layer is present for the mixture in the one-phase regime and it is defined as the difference between the real profile close to the surface $\phi(z)$ and the flat profile kept at bulk concen-tration ϕ_∞. It is described by the surface excess z*:

$$z^* = \int_0^{z(\phi_\infty)} [\phi(z) - \phi_\infty] dz \tag{23}$$

Here $z(\phi_\infty)$ is the distance from the surface (at depth z=0) to the plateau in composition. The surface *enriched/depleted* in blend component A is character-ized by positive/negative z* (see Fig. 15). Relatively large correlation lengths for polymer mixtures (see Sects. 2.1 and 2.2.2) lead to the surface profiles $\phi(z)$ of sufficient spatial extent that may be easily traced by current depth profiling tech-niques [29]. Surface enrichment has been observed at a free surface [164, 165] and at a substrate [92] as well as at an interface between binary blend and a homopolymer [166].

Already the original Cahn theory [153] finds the behavior of the surface excess layer observed in the one phase region to be related to the wetting phenomena. A logarithmic divergence of this excess layer is predicted [153] above the
wetting temperature T_w and for bulk concentration approaching the binodal val-
ue $\phi_\infty \to \phi_1$; this seems to be obeyed by segregation isotherms $z^*(\phi_\infty)$ determined
by us recently [16]. A jump in z^* from a small to a large, but finite value is predicted [153] to accompany (along a *pre-wetting line* marked in the surface phase
diagram of Fig. 13) the first order wetting. So far only a first order transition in
surface excess was observed experimentally [167, 168], but with z^* jumping
from a finite large to small value when bulk concentration ϕ_∞ was increased.

The Cahn approach describing simple fluid mixtures has been adopted by a
mean field theory developed for polymer mixtures by Nakanishi and Pincus [61]
and Schmidt and Binder [15] and is presented in the next section. The mean field
theory and its various extensions [7] have been successfully used to describe
much of the experimental segregation isotherm $z^*(\phi_\infty)$ data obtained so far [16,
92, 120, 145, 165–167, 169–175]. It allows us not only to distinguish isotherms
$z^*(\phi_\infty)$ characteristic for partial and complete (first and second order) wetting
but also to determine surface free energy parameters useful in predicting surface phase diagrams for the studied mixtures.

In recent work Jerry and Dutta [176] reanalyzed, with the mean field approach, conditions [8] of the second order wetting transition. They have found
that critical wetting transition must be accompanied by a prerequisite phenomenon of an *enrichment-depletion duality*; it is expected that the surface is enriched in the given component when bulk composition ϕ_∞ is below a certain value Q and is depleted in the same component for $\phi_\infty > Q$. Such an effect, easily predicted by simple lattice theory [177] and observed in Monte Carlo simulations
[178, 179], has been very recently determined by us for a real polymer blend
[175] (see Sect. 3.1.2.4).

Real polymer mixtures studied here do not form semi-infinite systems but
rather they are confined in thin layers bounded by two surfaces. For relatively
thick films (for a "critical" thickness evaluation see Sect. 3.2) the equilibrium
profile $\phi(z)$ of the whole film is described separately for each of the two surfaces
allowing their independent characterization. This is based on the assumption
that the profile $\phi(z)$, describing the segregation to the respective surface, is in
equilibrium with the plateau value of ϕ in the region adjacent to this surface.
This approach was justified by theoretical [180, 181] and experimental [182]
works on the dynamics of surface segregation, and is used here to focus on phenomena occurring near a single surface.

3.1.1
Mean Field Theory

For the case of a polymer mixture in contact with a surface ("external" interface)
at depth z=0, similar considerations apply [15, 61, 153] to those presented for the
interface separating two polymer phases (see Sect. 2.1 and discussion following

Eq. 7). Now the overall free energy F (per site volume Ω and area A normal to the surface) consists of a bulk "gradient square" functional (strictly valid in a long wavelength limit) and a specific "bare" surface contribution $f_s(\phi_s)$. Bulk contribution is now summed over the half-space (z>0) from the surface:

$$\frac{F}{k_BTA/\Omega} = f_s(\phi_s) + \int_0^\infty dz\{\Delta F_M[\phi(z)] - \Delta\mu\phi(z) + \kappa[\nabla\phi(z)]^2\} \tag{24}$$

The "bare" surface contribution $f_s(\phi_s)$ is assumed [15, 61, 153] to be *short ranged* in the sense that it is a function of the surface concentration ϕ_s. In some of the current theories [7, 183–186] f_s also includes surface concentration gradient terms reformulated by experimenters [120, 167] to obtain f_s expressions dependent only on ϕ_s.

A standard variational calculus, extended slightly as compared to that described in Sect. 2.1, is applied to find the profile $\phi(z)$ and its surface value, minimizing Eq. (24). It yields a differential equation describing the profile $\phi(z)$ (identical to Eq. 8b):

$$-2\kappa(\phi)\frac{d\phi}{dz}(\phi) = \pm 2\sqrt{\kappa(\phi)\Delta f(\phi;\chi,\phi_\infty)} \tag{25}$$

along with an additional surface boundary condition:

$$\frac{-df_s}{d\phi}(\phi_s) = -2\kappa\left(\frac{d\phi}{dz}\right)_s = \pm 2\sqrt{\kappa(\phi)\Delta f(\phi_s;\chi,\phi_\infty)} \tag{26}$$

Here the parameter Δf, completely determined by bulk parameters χ and ϕ_∞ (Eq. 8a), describes the energy needed to create a (local) unit volume of a blend with composition $\phi(z)$ from a bulk reservoir kept at composition ϕ_∞. The profile $\phi(z)$, given by Eq. (25), is cut off at the surface by the surface concentration ϕ_s specified by the condition at Eq. (26). The gradient of the profile from the surface (at z=0) to the bulk (z>0) may be *negative* or *positive* corresponding to the surface *enrichment* or the surface *depletion*, respectively (see Fig. 15). To distinguish these two cases, we always use the convention in which the *upper* sign stands for the *negative* and the *lower* one for the *positive* concentration gradient.

The total excess free energy F_e of the surface excess layer, defined as the difference between the energy F (of Eq. 24) necessary to build up the equilibrium profile $\phi(z)$ and the energy F of the flat profile kept at bulk composition ϕ_∞, is equal to

$$\frac{F_e}{k_BTA/\Omega} = f_s(\phi_\infty) - \int_{\phi_\infty}^{\phi_s} [\frac{-df_s}{d\phi} \mp 2\sqrt{\kappa}\ \sqrt{\Delta f}]d\phi \tag{27}$$

In his original work, Cahn [153] has introduced a graphical technique, later called the *"Cahn construction"*, to find all allowed profiles $\phi(z)$ solving Eqs. (25)

and (26), and to evaluate graphically their corresponding excess energies F_e defined by Eq. (27). He has plotted $-2\kappa\nabla\phi$ vs the composition ϕ and $(-df_s/d\phi)_s$ vs ϕ_s in the same set of axes. The relation $-2\kappa\nabla\phi(\phi)$ is topologically equivalent to the "phase portrait" (introduced by Fig. 2); from it the local curvature as well as the whole profile $\phi(z)$ may easily be deduced. Here it represents all profiles allowed by Eq. (25) for given bulk parameters (see solid and dashed lines in Fig. 15). The function of the surface energy derivative $(-df_s/d\phi)_s$ vs ϕ is marked by dotted lines in Fig. 15. At each intersection point of these two relations the surface boundary condition (Eq. 26) is clearly met and specified by the related surface concentration ϕ_s ($=\phi_{se}$ or ϕ_{sd} in Fig. 15). Additionally, the reduction in the excess free energy F_e, associated with the profile $\phi(z)$ expanding from ϕ_s to ϕ_∞, is easily represented by the corresponding area S (shaded regions in Fig. 15). In Fig. 15 we have presented two Cahn constructions, typical for the case of surface enrichment and surface depletion.

The graphical Cahn technique [8, 61, 153, 176] and its numerical analogs [15, 159, 162, 187] have been used to study the conditions for first and second order wetting transition. In these calculations two simplifying assumptions, not valid in real polymer blends [16, 120, 167, 170, 175] (see the following sections), have been imposed on the "bare" surface contribution f_s; it is taken to be temperature independent and is approximated by the first two terms of an expansion in the volume fraction at the surface ϕ_s:

$$-f_s(\phi_s) = \mu_1\phi_s + \frac{1}{2}g\phi_s^2 \tag{28}$$

The Cahn constructions used are presented in Fig. 14. The allowed trajectories $-2\kappa\nabla\phi(\phi)$ are marked by $+2(\kappa\Delta f)^{1/2}$ and $-2(\kappa\Delta f)^{1/2}$ curves plotted for bulk composition equal to a binodal one $\phi_\infty = \phi_1$. They are equal to zero at coexistence compositions ϕ_1 and ϕ_2 and have extrema at a concentration close to the critical value ϕ_c. Both binodal values are shifting towards ϕ_c as temperature is increased. Simultaneously, the width of the interface between ϕ_1 and ϕ_2 increases (see Figs. 2 and 7) leading to smaller humps in $+2(\kappa\Delta f)^{1/2}$ and $-2(\kappa\Delta f)^{1/2}$. The temperature independent surface energy derivative $(-df_s/d\phi)_s$, corresponding to Eq. (28), is represented in Fig. 14 by a straight dotted line.

The first order wetting transition may occur if $(-df_s/d\phi)_s$ does not change its sign when ϕ is varied. Then $(-df_s/d\phi)_s$ may intersect the trajectory $-2\kappa\nabla\phi(\phi)$ in a specific way depicted in Fig. 14b. Only two, out of four, intersection points – ϕ_{1s} and ϕ_{2s} – correspond to locally stable solutions of the variational problem. They describe two surface excess layers (see Fig. 14a) exhibiting partial ($\phi_1 < \phi_{1s} < \phi_2$) and complete wetting ($\phi_1 < \phi_2 < \phi_{2s}$), respectively. The excess free energy F_e of these two composition profiles may be calculated with Eq. (27). Their energy differs by ΔF_e presented in terms of the areas S_1 and S_2 in Fig. 14b:

$$\frac{\Delta F_e}{A} = \frac{F_e(\phi_{s2}) - F_e(\phi_{s1})}{A} = \frac{k_B T}{\Omega}(S_1 - S_2) \tag{29}$$

and related [8] to the spreading coefficient $(\gamma_{1s}-\gamma_{2s}-\gamma_{12})=-\Delta F_e/A$. For temperatures $T<T_w$ the area S_1 is larger than S_2 and a partial wetting solution corresponds to the lowest energy. The opposite is true at $T>T_w$ where complete wetting is expected. At the wetting point T_w both areas S_1 and S_2 are equal and the discontinuous transition between both solutions occurs. A construction similar to the presented here may be drawn for the pre-wetting transition [153].

Substantially different behavior is predicted [176] when $(-df_s/d\phi)_s$ changes its sign at concentration Q as it is shown in Fig. 14c. At temperatures below T_w the surface energy derivative $(-df_s/d\phi)_s$ intersects trajectories $-2\kappa\nabla\phi(\phi)$ at two points – ϕ_{1s} and ϕ_{2s}. The free energy of these two solutions differs by ΔF_e expressed using Eq. (29) in terms of the area S_1 shown in Fig. 14c ($S_2=0$). Thus whenever $\phi_1<Q<\phi_2$ the solution with minimal energy F_e would always correspond to a partial wetting ($\phi_1<\phi_{1s}<\phi_2$). With increasing temperature the humps in trajectories $-2\kappa\nabla\phi(\phi)$ become smaller and ϕ_{1s} increases continuously with respect to upper binodal ϕ_2. Finally at T_w the intersection point ϕ_{1s} reaches ϕ_2 at the concentration Q and the complete wetting regime is attained. The basic criterion for the critical wetting – $0<Q<1$ – is also a condition for the enrichment-depletion duality to occur (consider Fig. 15 for $Q_e=Q_d$). Thus the duality is a prerequisite of critical wetting [176].

3.1.2
Review of Experimental Results

3.1.2.1
Surface Segregation

The profile $\phi(z)$ of the surface segregated layer is predicted by a mean field theory to follow the formula (obtained by integrating Eq. 25):

$$z(\phi)=\mp\int_{\phi_s}^{\phi(z)}\sqrt{\kappa(\phi)/\Delta f(\phi;\chi,\phi_\infty)}d\phi \qquad (30)$$

It is described entirely by two bulk (χ and ϕ_∞) and one surface (ϕ_s) parameters. With surface concentration ϕ_s being close to the bulk value ϕ_∞ the parameter $\Delta f(\phi;\chi,\phi_\infty)$ may be approximated by the bulk correlation length ξ (Eq. 6): $\Delta f(\phi;\chi,\phi_\infty)\approx\xi^{-2}(\phi_\infty)\kappa(\phi_\infty)(\phi-\phi_\infty)^2$. In such a limit (valid strictly for $\phi_\infty<\phi_1$) an exponential behavior of the surface profile is obtained:

$$\phi(z)=\phi\infty+(\phi s-\phi\infty)\exp(-z/\xi) \qquad (31)$$

This formula may also well approximate the strict solution (Eq. 30) for larger values of bulk concentration ϕ_∞ but in this case its decay length is no longer equal to the bulk correlation length ξ. The mean field theory also makes a prediction for the total surface excess z*:

$$z^*(\phi_\infty)=\pm\int_{\phi_\infty}^{\phi_s}(\phi(z)-\phi_\infty)\sqrt{\kappa(\phi)/\Delta f(\phi;\chi,\phi_\infty)}d\phi \qquad (32)$$

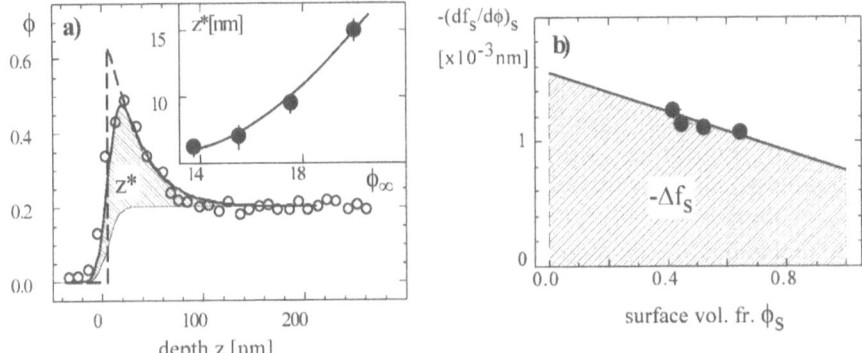

Fig. 16.a The free surface region of the isotopic polystyrene mixture with binodal line presented in Fig. 7 (with $\chi=(.124/T -1.06\times10^{-4})(1-.18\phi)$) after 30 days of annealing at T= 170 °C: Measured [92] composition-depth $\phi(z)$ profile (○) is compared with that predicted by the mean field theory before (*dashed line*) and after (*solid line*) convolution with the resolution of the experimental method ($\rho\approx12$ nm HWHM). **b** Results of the Cahn construction done for the segregation data of the isotherm T=170 °C (see the inset to a) and corresponding to the profile of a. Calculated composition derivatives of "bare" surface free energy (−$df_s/d\phi)_s$ (● points) are fitted by a *solid line* with $\mu_1=1.6\times10^{-3}$ nm and g=-7.6×10^{-4} nm. The *hatched area* is related to the surface energy difference −Δf_s

dependent, again, on two bulk (χ and ϕ_∞) and one surface (ϕ_s) parameters.

All *direct* depth profiling techniques used to study the surface segregation from binary polymer mixtures have a depth resolution [29] ρ limited to some 5–40 nm HWHM (half width at half maximum of the related Gaussian function). They cannot observe the real composition profile $\phi(z)$ (for the sake of comparison mimicked by mean field prediction (dashed line) in Fig. 16a) but rather its convolution (solid line in Fig. 16a) with an instrumental resolution function characterized by ρ. The total surface excess z* however provides a good parameter, independent of resolution, as it has been concluded based on experimental data obtained using different direct techniques [170].

Thus, in a commonly used procedure [16, 92, 120, 145, 165, 167, 170, 175], each point of the segregation isotherm data $z^*(\phi_\infty)$ is analyzed with Eq. (32) to find out the surface concentration $\phi_s(\phi_\infty)$ assigned by the mean field approach (analogous procedure exists in a self-consistent mean field model [166, 174]). Then, for each such pair (ϕ_∞, ϕ_s) the trajectory $\pm2(\kappa\Delta f)^{1/2}$ vs ϕ is plotted (for the value ϕ_∞) and its value is read out at surface concentration ϕ_s (=ϕ_{se} or ϕ_{sd} in Fig. 15). This value is equal to the surface energy derivative (−$df_s/d\phi)_s$ at concentration ϕ_s. Such a procedure, repeated for each $z^*(\phi_\infty)$ data point, yields the concentration dependence of the composition derivative of the short-ranged "bare" surface energy (−$df_s/d\phi)_s$ vs ϕ_s (see Fig. 16).

Numerous trials for the mean field approach have been performed. Most of them have been focused on testing whether real profiles $\phi(z)$ of the surface segregated layer follow Eq. (30). Neutron reflectometry (NR, with a depth resolution of ca.1 nm) has been used particularly often. Although it is an *indirect* method and its data analysis depends strongly on the chosen model for $\phi(z)$, it is extremely sensitive (± 0.01) to the volume fraction ϕ_s at the surface regardless of the detailed functional form of the profile $\phi(z)$ [29, 169]. Two recent NR experiments [169, 188] have found small but experimentally significant profile shape deviations from the Eq. (30) form, indicating that there may be a flattening of the profile at the vacuum surface over a distance of about 10% and 30–50% of the bulk correlation length for the isotopic mixtures of polystyrene (PS) [169, 170] and poly(ethylene propylene) (PEP) [188], respectively. A similar phenomenon has also been observed in a recent Monte Carlo simulation [179]. Such a significant flattening has not been observed, however, in other NR [173, 174] or NRA [189] measurements.

A theory has been developed by Chen et al. [190] to explain the profile flattening. They assumed that the "bare" surface energy is not of a short-ranged character $f_s(\phi_s)=f_s(\phi)\delta(z)$ but is rather dominated by a *long-ranged* contribution $v(z)\propto 1/z^3$. The contribution $v(z)$ may be related [191] to van der Waals forces: $v(z)=-\Delta A/(6\pi z^3)$, where ΔA is the difference in Hamaker constants between two blend components (expressed by their surface tension difference $\Delta\gamma$ [191]). The overall free energy, accounting now only for van der Waals surface interactions, is given by the modified expression (Eq. 24) [190, 191]:

$$\frac{F}{k_B TA/\Omega} = \int_0^\infty dz \{\Delta F_M[\phi(z)] - \Delta\mu\phi(z) + \kappa(\phi)[\nabla\phi(z)]^2 - v(z)\phi\} \qquad (33)$$

Variational calculus leads now to a rather non-trivial equivalent of Eq. (25). For instance the surface enrichment for a partial wetting is no longer described by a part of interfacial profile cut off at the surface. In turn, the surface boundary condition (left and central part of Eq. 26) yields directly, for the absent short range contribution f_s, the zero composition gradient at the surface ($d\phi/dz)_s$. For polymer blends the profile $\phi(z)$ minimizing Eq. (33) was found [191], however, to be flat only at a 10^{-1} nm deep surface region and to be otherwise *indistinguishable* from the contour $\phi(z)$ solving the standard expression (Eq. 24) for identical surface tension difference $\Delta\gamma$ assumed (related to f_s with Eq. 34). Self-consistent mean filed calculations [174] lead to similar conclusions. Therefore, while long range surface forces are nowadays believed to be in charge of wetting dynamics [160, 192] (related to surface segregation phenomena), the computationally much simpler short-range f_s approximation is used instead, with an equal confidence, to calculate the equilibrium profiles $\phi(z)$.

Observations by NR profile flattening are interpreted in terms of deficiency in the mean field theory. However the ambiguity in fitting NR data may in principle weaken this conclusion. To make conclusions regarding the mean field theory more clear, another approach has been initiated [169, 173, 174] in which two complementary methods, i.e., direct depth profiling (yielding the overall profile) and indirect NR (providing superior resolution), are applied to the same system. Such extended comparison of both complementary methods [167, 173] might

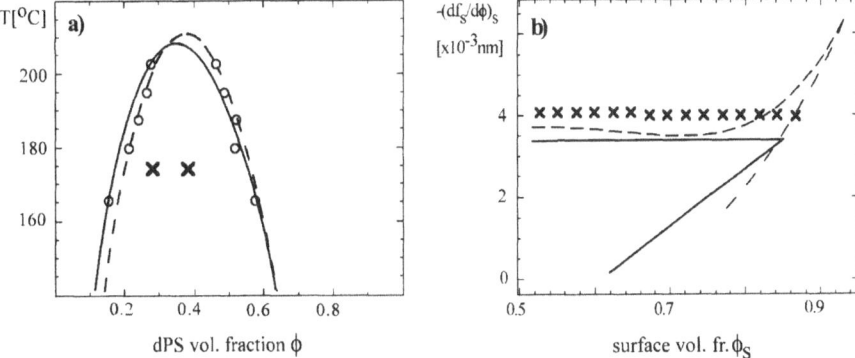

Fig. 17.a Coexistence curve of the dPS (N_{dPS}=5714)/PBr$_{0.062}$S (N_{PBrxS}=1660) mixture, determined (O points) by [167] and described by χ=.25/T+1.94×10^{-4} (*solid line*) [167]) and by χ=(.197/T+3.03×10^{-4}) (1+.059ϕ) (*dashed line*). Coexistence concentrations of the very similar dPS (N_{dPS}=6205)/PBr$_{0.06}$S (N_{PBrxS}=1587) mixture (X points) described [173] at T= 174 °C by χ=7.171×10^{-4}. **b** The composition derivatives of the "bare" surface free energy (–df$_s$/dϕ)$_s$ vs ϕ calculated for the segregation isotherm z*(ϕ_∞) reported by [167] for the dPS (N_{dPS}=5714)/PBr$_{0.062}$S (N_{PBrxS}=1660) mixture at T=180 °C (*solid* [167] *and dashed curves*) and evaluated [173] based on the z*(ϕ_∞) data for the very similar dPS (N_{dPS}=6205)/PBr$_{0.06}$S (N_{PBrxS}=1587) mixture at T=174 °C (X points). Each of the various results of the Cahn construction presented in b is obtained for the assumed specific coexistence conditions (described by different χ forms) in a. The results in b and corresponding coexistence conditions in a have the same denotations

be, for instance, made for a deuterated polystyrene dPS/poly (styrene-4-*co*-bromostyrene) PBr$_x$S system. Forward recoil spectrometry (FRES) studies [167] of the dPS (m.w.=6.4×10^5)/PBr$_{0.062}$S (m.w.=1.7×10^5) mixture with dPS segregating towards the vacuum interface, indicated at two temperatures (180 °C and 200 °C), the jump in the segregation isotherm z*(ϕ_∞), from a finite larger to a smaller value, interpreted in terms of two branches of the surface energy derivative (–df$_s$/dϕ)$_s$ on the related Cahn construction (see solid lines in Fig. 17b). On the other hand, NR measurements [173] of the free surface of the almost identical dPS (m.w.=6.95×10^5)/PBr$_{0.06}$S (made from PS with m.w.=1.65×10^5) mixture have shown [173] a monotonic segregation isotherm z*(ϕ_∞) at T=174 °C and a single linear (–df$_s$/dϕ)$_s$ vs ϕ_s relation on the Cahn plot (marked by X in Fig. 17b). Both measured isotherms are very similar and differ by the single z*(ϕ_∞) point preceding the jump in z* and corresponding to a metastable surface state. Metastable states have been observed a few times in segregation experiments [167, 168] but apparently not in a discussed NR run [173]. The most striking difference between the two reports lies in the results of performed Cahn constructions [167, 173] (Fig. 17). This disparity is caused mainly by the various values of the interaction parameters (differing by 5%) used: while the χ value used by the FRES group [167] describes the determined earlier coexistence curve, the χ value used in NR data analysis [173] was evaluated reasonably from other data but it was not related to the relevant coexistence curve (see Fig. 17). To support this conclusion we show here that even small changes in the χ form (associated by the change of$<\chi>$ by some 2%), made to improve the overall description of coexistence curve, lead to significant changes in the related Cahn construction (calculated dashed lines in Fig. 17). It is comprehensible and commonly accepted that the knowledge of the binodal line is essential to interpret properly an observed surface segregation. We emphasize here that it is even inevitable to determine experimentally the coexistence curve for the same mixture for which the surface phenomena are to be analyzed in a quantitative fashion. It is not enough to extrapolate the interaction parameter χ determined elsewhere. This strict approach has been accepted however by only two research groups [14, 16, 74, 75, 88, 91, 92, 94–97, 115, 117, 120, 145, 167, 175] (very recently also by a third one [98]).

An ideal series of experiments set to check the mean field theory, which would consist of the binodal determination and succeeding surface segregation studies with NR and with direct depth profiling techniques, is still missing. Genzer et al. [174] have studied surface segregation in another dPS/PBr$_x$S system by FRES and NR, yielding complementary data on bulk ϕ_∞ and surface ϕ_s concentrations of the same samples, but have used the χ parameter interpolated from a similar system with different bromination level x. The more exact (especially ϕ-dependent) form of the interaction parameter used, may in principle improve the good qualitative agreement between measured and predicted by the mean field: $z^*(\phi_\infty)$ and $\phi_s(\phi_\infty)$ relations concluded from their work. It has been shown [174] that this agreement may be further improved by use of the self consistent mean field (SCMF), which is not bounded by the long wavelength limit approximation of the mean field. This limit is certainly not valid for the steepest surface regions of the concentration profiles $\phi(z)$ measured at bulk concentrations ϕ_∞ far from binodal. However the profiles for bulk concentrations close to the binodal have slopes not exceeding by much those characteristic for the coexisting phases where the long wave limit has been shown to be obeyed (see Sect. 2.2.2). The earlier work by Genzer et al. [193], also focusing on testing mean field and SCMF predictions based on data by Zhao et al. [170], seems to be controversial as the ϕ_s values used were not determined directly (e.g., by NR) but were fitted [170] to profiles yielded by a profiling method (SIMS) with a limited resolution ρ. A completely different approach has been used by Cifra et al. [194] who found that the mean field approach gives an adequate description of the compressible polymer blend simulated in Monte Carlo study.

3.1.2.2
"Bare" Surface Free Energy

A *surface energy difference* Δf_s and (a *surface tension difference* $\Delta\gamma$) between two pure blend components may be evaluated based on a $(-df_s/d\phi)_s$ vs ϕ_s relation calculated for measured segregation isotherm data:

$$\Delta f_s := f_s(\phi_s = 1) - f_s(\phi_s = 0) = -\int_0^1 (\frac{-df_s}{d\phi})_s d\phi, \qquad \Delta\gamma = \frac{k_B T}{\Omega} \Delta f_s \qquad (34)$$

The surface energy difference may also be expressed by a dimensionless parameter χ_s defined per lattice site. With a lattice site size taken as $\Omega^{1/3}$ and the corresponding surface area $A=\Omega^{2/3}$ occupied by one lattice site we may relate the *surface energy difference parameter* χ_s with Δf_s as (see Eq. 24):

$$\chi_s=(A/\Omega)\,\Delta f_s=\Delta f_s/\Omega^{1/3} \tag{35}$$

For most practical polymer blends the surface tension differences $\Delta\gamma$ are usually greater [195] than 2 mJ/m². Much lower values are obtained for isotopic mixtures. For the isotopic mixture of polystyrene hPS (m.w.=2.89×10⁶)/dPS (m.w.=1.03×10⁶) (see Figs. 7 and 16) we have determined [92] the surface tension at a free surface to be reduced at T=170 °C for the deuterated component by $\Delta\gamma$=4.2×10⁻² mJ/m² as compared to protonanted PS (described equivalently by Δf_s=−1.22×10⁻³ nm or by χ_s=−2.1×10⁻³). The comparison with other evaluations is only in moderate agreement: the data by Steiner [88] yield χ_s=−2.9×10⁻³ for hPS (m.w.=1.8×10⁶)/dPS (m.w.=1.95×10⁶) at 218 °C while the value χ_s=−3.7×10⁻³ concluded based on the pioneer work by Jones et al. [165] for hPS (m.w.= 1.8×10⁶)/dPS (m.w.=1.03×10⁶) at 184 °C had been changed into χ_s=−4.6×10⁻³ after the supplementary part of the segregation isotherm z*(ϕ_∞) at higher bulk concentrations was measured [170]. While one possible reason for such discrepancies is the usually limited set of $(-df_s/d\phi)_s$ vs ϕ_s points used to evaluate Δf_s (see Fig. 16), the other one is related to different forms of the interaction parameter χ used in Cahn constructions.

3.1.2.2.1
Enthalpic Contributions to "Bare" Surface Free Energy f_s

The driving force for the surface segregation is most commonly interpreted in terms of *enthalpy*-based arguments developed by a simple mean field model [177, 196]. Even in the simplest case of the mixture bounded by a *free (neutral) surface* (modeled by a vacuum//blend "external" interface) the contribution to the short-ranged "bare" surface energy f_s is expected. The reason for this is the removal of the mixture from the region above the surface. As a result half of the interactions executed by polymer segments located at the surface is lost. Hence the internal energy E_{int} is reduced by one half and the *missing neighbor* contribution to surface energy $f_s^{H,mn}$ may be expressed as [177]:

$$f_s^{H,mn}=\frac{-1}{2}E_{int}=\frac{-(z_s/\Omega^{2/3})\Omega}{2k_BT}[\phi_s^2\varepsilon_{AA}+(1-\phi_s)^2\varepsilon_{BB}+2\phi_s(1-\phi_s)\varepsilon_{AB}] \tag{36}$$

where $\Omega^{2/3}$ stands for the surface area A occupied by one lattice site and z_s is the surface coordination number, i.e., the number of surface/mixture contacts per lattice site. We notice the quadratic in ϕ_s form of f_s as postulated by Eq. (28) with non-zero both μ_1 and g coefficients.

The corresponding surface energy difference parameter $\chi_s^{H,mn}$ may be expressed (see Eq. 35) in terms of the contact energies ε_{ii} or the solubility parameters δ_i of blend components (see Eq. 16):

$$\chi_s^{H,mn} = \frac{-1}{2} \frac{z_s}{k_B T}[\varepsilon_{AA} - \varepsilon_{BB}] = \frac{z_s}{z_b - 2} \frac{V}{k_B T}[\delta_A^2 - \delta_B^2] \tag{37}$$

Here the lattice site volume Ω is taken to be equal to the average segmental volume V of both blend components. Equation (37) predicts that a blend component with a lower cohesive energy density would have a lower surface tension. When the contact energies ε_{ii} or the solubility parameters δ_i are known from bulk measurements, the evaluation of the surface interaction parameter due to the missing neighbor effect $\chi_s^{H,mn}$ is feasible [145, 197]. The value $\chi_s^{H,mn}=-5\times10^{-3}$, calculated by Kumar and Russell [197] for the isotopic polystyrene mixture, is in reasonable agreement with the above-mentioned experimental χ_s values obtained for the segregation to the free surface.

The *real surface* (e.g., the substrate//mixture interface) may prefer one of the blend components. This is expressed by different contact interactions between surface (S) and both blend component (A, B) molecules $\varepsilon_{SA} \neq \varepsilon_{SB}$. The resulting surface energy f_s^H is

$$f_s^H = \frac{(z_s/\Omega^{2/3})\Omega}{k_B T}[\phi_s \varepsilon_{SA} + (1-\phi_s)\varepsilon_{SB}] - \frac{1}{2}E_{int} \tag{38}$$

The "quadratic in ϕ_s' form is preserved with μ_1 – originating now mainly due to the difference in surface-polymer contact energies and g – entirely specified by missing neighbor effect. Both coefficients appearing in the linear form of the surface energy derivative $(-df_s/d\phi)_s=\mu_1+g\phi_s$ may be expressed as [177]

$$\mu_1 = \frac{2z_s}{z_b - 2}\Omega^{1/3}\chi Q \qquad g = \frac{-2z_s}{z_b - 2}\Omega^{1/3}\chi \tag{39}$$

The surface concentration Q at which $(-df_s/d\phi)_s$ changes sign is related to the hypothetical interaction parameters between surface and blend component molecules χ_{SX} (X=A or B), defined similarly to FH interaction parameter χ (Eq. 2):

$$Q = \frac{1}{2}(1 + \frac{\chi_{SB} - \chi_{SA}}{\chi}), \qquad \chi_{SX} = \frac{z_b - 2}{k_B T}(\varepsilon_{SX} - \frac{1}{2}(\varepsilon_{SS} + \varepsilon_{XX})) \tag{40}$$

Both surface free energy parameters μ_1 and g are found to be χ, and hence also temperature, dependent. This fact has also been noticed by others [99, 159, 178, 194, 196]. While for the real surface $Q \gg 1$ (or $Q \ll 0$) as χ_{SA} differs from χ_{SB}, for the neutral (free) surface $0 < Q < 1$ may be obtained for blend components with similar cohesive energy. Then the surface enrichment/depletion occurs, as observed in Monte Carlo simulations performed for mixtures with $\varepsilon_{AA}=\varepsilon_{BB}(\neq\varepsilon_{AB})$ and the resulting concentration Q=1/2 [178, 179]. When the real surface is composed of the immobilized polymer mixture kept at concentration ϕ_{fw} ("*frozen wall*" case [177]) then Eq. (39) would be still valid for Q=ϕ_{fw}. Experimental studies of situations close to the frozen wall case have been initiated [166].

An extension of the short range effects beyond nearest neighbors leads [185] to concentration gradient terms $[(d^n\phi/dz^n)_s]^m$ being present in quadratic f_s form of Eq. (28).

3.1.2.2.2
Entropic Contributions to "Bare" Surface Free Energy f_s

The Cahn analyses already performed for early surface enrichment observations [167, 170] have shown that the yielded relation of the surface energy derivative $(-df_s/d\phi)_s$ vs ϕ_s cannot be described by the linear form suggested by simple arguments presented above. While it can be argued that the more sophisticated enthalpy-based models (e.g., [185]) might eventually account for these discrepancies, the relation $(-df_s/d\phi)_s$ has been interpreted with the Cohen and Muthukumar model [183] instead. This model considers *entropic* effects due to the restriction of the configuration of polymers in the vicinity of an "external" interface (surface) and finds the additional entropic contribution to the "bare" surface free energy f_s^S to be equal to

$$f_s^S = \alpha_1\phi_s\ln\phi_s + \beta_1(1-\phi_s)\ln(1-\phi_s) + [\alpha_2\ln\phi_s + \beta_2\ln(1-\phi_s)](\frac{d\phi}{dz})_s \quad (41)$$

α_i and β_i terms being complicated functions of binding potentials between the surface and the segments of the respective species [183].

The first surface segregation experiment, which cannot be explained even qualitatively by simple enthalpic considerations, has been reported by Hariharan et al. [171]. They studied isotopic polystyrene (PS) blends with different degrees of polymerization (N_{hPS}, N_{dPS}) of both components. They observed that while the surface is enriched in deuterated PS for comparable N_{dPS} and N_{hPS} with $\chi_s\approx-5\times10^{-3}$, the protonated PS component is preferred at the surface for $N_{hPS}<<N_{dPS}$ with corresponding $0<\chi_s<7\times10^{-2}$. This reversal of the isotopic surface enrichment is attributed to an additional entropic driving force preferring shorter polymer chains at the surface. The related contribution to the surface free energy is due to different density gradients of both polymer blend components at the surface. Polymer density gradient present at the surface, but neglected by constant density lattice models, is shown to be dependent on polymer molecular weight [171].

The surface segregation from the mixture of chemically identical polymers with chain length disparity is predicted by another model [198–200]. It represents the spatial conformation of a polymer coil as a random walk reflected by an "external" interface. The associated loss in system configurations is minimized when shorter chains are adsorbed at the surface. Preferential surface segregation of comb-like additives from a linear polymer matrix is similarly predicted [199, 201] and observed [202]: here long-branched additives act as a collection of unconnected short linear chains.

Another experiment with a much larger impact has been reported by Sikka et al. [203]. They have studied lamellar structures created in thin films by diblock copolymers, composed of two blocks with similar elementary chemical units but with different statistical segment length. Their results show that the lamellae of the block with a shorter segment length are always formed at an "external" interface independent of its nature. It was suggested that an entropy driven mechanism, postulated by Fredrickson and Donley [184], would explain the observed surface segregation. The Fredrickson-Donley model considers the mixture composed of two homopolymers with different statistical segment lengths a_i and a_j. The entropic effects related to the polymer configurations perturbed by the surface lead to an additional surface energy term, written as

$$f_s^S = \frac{-1}{12}[a_i^2 - a_j^2](\frac{d\phi}{dz})_s \tag{42}$$

which tends to favor an enrichment of the components with shorter segment length. In the later version [186] of this theory the parameter $\beta_i = a_i/(6V)^{1/2}$ stands for a_i in Eq. (42). Another work [204] suggests that the Fredrickson-Donley model may not apply to the results obtained for diblocks. Very recently, Fredrickson et al. have suggested an interplay between entropic and enthalpic factors driving the segregation [205, 206]. The results by Sikka et al. [203] have stimulated much of the theoretical works [184, 186, 204–209] on homopolymer mixtures with statistical segment length disparity and have heightened an interest in experimental studies [14, 16, 115, 117, 120, 145, 160, 175, 210–212] of their real counterparts.

The entropic contribution to the surface free energy f_s^S would be of importance for binary mixtures with a small chemical mismatch between their components, so that the surface energy differences of enthalpic origin f_s^H are minimized (consider Eq. 39 for a small χ). Polyolefine blends have been proposed as optimal candidates to test f_s^S in the original work of Fredrickson and Donley [184]. The mixtures studied in detail (see Sects. 3.1.2.3–3.1.2.5 and also Sect. 2.2.3) consist of the statistical olefinic copolymers of structure $(E_{1-x}EE_x)_N$; where linear (C_4H_8) ethylene (E) and branched $(C_2H_3(C_2H_5))$ ethyl ethylene groups (EE) are distributed at random with relative frequencies $(1-x):x$ along the chain backbone (see Fig. 18). They may be regarded as "effective homopolymers" whose mean microstructure $E_{1-x}EE_x$ varies continuously with x from polyethylene (x=0) to poly(ethyl ethylene) (x=1). The change in microstructure x involves the variation of statistical segment length a_x: the more branched the chains (i.e., with higher x) the shorter the segment lengths [120]. This drastic change in a_x $((a_0-a_1)/<a>\approx 50\%)$ is not followed by segmental volume $((V_0-V_1)/<V>\approx 1\%)$ [141, 142]. Recently, a mapping has been proposed relating the real hydrocarbons with the chains representing them in coarse-grained models

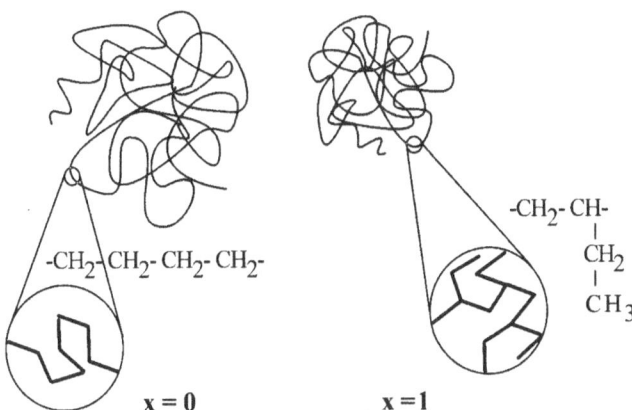

Fig. 18. Two isomers of the random copolymer $(E_{1-x}EE_x)_N$: a stiff molecule of linear polyethylene $E_N=(C_4H_8)_N$ (x=0) with a larger statistical segment length a and a flexible chain of branched poly(ethyl ethylene) $EE_N=(C_2H_3(C_2H_5))_N$ (x=1) with a smaller statistical segment length a

[204, 207, 209]. An aspect ratio $\Gamma=a/V^{1/3}$ has been introduced [213], characterizing an effective stiffness of the chain and reflecting an interchain packing ability. The real more *linear* (low x) olefinic copolymers are described as the more *stiff* chains (with large Γ and a), while the more *branched* polyolefines (high x) as the more *flexible* blend components (with small Γ and a).

Recent results of Monte Carlo simulations and integral equation theory by Kumar and co-workers [207, 209] suggest that the *configurational entropy* effect, advocated by earlier theories [183, 184, 186], competes with *packing entropy* and entropy effects due to local rearrangements of segments at the surface. The last two entropic contributions, favoring stiff chains at the surface, are found to prevail in mixtures with stiffness disparity at melt-like polymer densities [207, 209]. Recent mean field studies also indicate the surface excess of the more rigid macromolecules [200, 204, 214]. In another Monte Carlo simulation [208], in which coarse-grained chains still have distinct linear or branched topology, a small surface enrichment in the linear (i.e., stiff) component is observed but only for the vanishing fluid-fluid attractions. Much larger segregation of branched chains is concluded for non-zero fluid-fluid attractions.

The divergent predictions of theories on entropic surface segregation advocating the surface enrichment in more flexible [184, 186, 205, 206], more stiff [200, 204, 207, 209, 214], or both types of molecules [183, 208] may be compared with experimental results obtained for blends composed of the statistical olefines $E_{1-x}EE_x$. The more branched (i.e., more flexible) component was found preferred at the free surface (with an exception of enrichment-depletion duality described in Sect. 3.1.2.4) but no segregation was observed at silicon or gold interfaces. This leads to the conclusion [120] that the entropic driving force alone cannot be in charge of the enrichment because in such a case the enrichment should be visible at all surface types. Instead of the above, the enthalpy-based explanation for the surface enrichment at the free surface has been suggested [120], which relates the segregation to the lower cohesive energy of the more flexible component (see Eq. 37 and Sect. 2.2.3). The more detailed quantitative analysis [145] of all available free surface segregation data [16, 120, 175], as presented in Sect. 3.1.2.5, leads to a more complicated picture involving both enthalpic and entropic driving forces.

3.1.2.3
Temperature Dependence of Surface Segregation and Wetting

To get the clear picture of surface phase diagram it is necessary to study segregation and wetting phenomena at different temperatures. Initially such studies have not been performed due to very low mobility of the high molecular weight polymers used [92, 165, 169, 170, 172] and hence extremely long times necessary to reach equilibrium state in annealing experiments. Recently, two types of blends have been used with more mobile molecular components. The dPS/PBr_xS system has been investigated focusing on the first order surface transition at two different temperatures [167]. More extended temperature studies of this system have been performed only very recently [99, 168]. The most detailed, so far, research on the role of temperature on the surface enrichment and wetting has been performed by us [16, 120, 145]. We have investigated the free surface of eight various binary blends dx_1/hx_2 composed of random olefinic copolymers $E_{1-x}EE_x$ with different composition $x_1 \neq x_2$ and different isotopic status (partially deuterated chains are denoted by "d" and their fully protonated analogs by "h"). These mixtures were grouped in four blend pairs, dx_1/hx_2 and hx_1/dx_2, with exchanged deuterium labeled component. Our primary interest, discussed in this section, was to evaluate the surface phase diagrams for the blends studied. The more general goal was to extract the surface energy difference parameter χ_s at the same reference temperature ($T_{ref}=100\,°C$). This allowed us to relate the change in χ_s upon the swapping of the deuterated blend component (see Sect. 3.1.2.5) with a similar effect observed for bulk interaction parameter χ (see Sect. 2.2.3).

Two different types of behavior characterize the temperature dependence of surface segregation in blends composed of the olefinic copolymers. Larger surface energy difference Δf_s between blend components (say at T_{ref}) results in complete wetting behavior observed even far below the critical point T_c. On the other hand, a small difference in Δf_s (at the same T_{ref}) makes the wetting transition possible, in principle, at temperatures close to T_c.

3.1.2.3.1
Blends With a Large Surface Energy Difference Δf_s

At least two of the studied blends [16, 145], d66 (degree of polymerization $N_{66}=2030$, deuteration extent e=0.4)/h52 ($N_{52}=1510$) and d86 ($N_{86}=1520$, e=0.4)/h75 ($N_{75}=1625$), may be described by a relatively large force driving to the surface

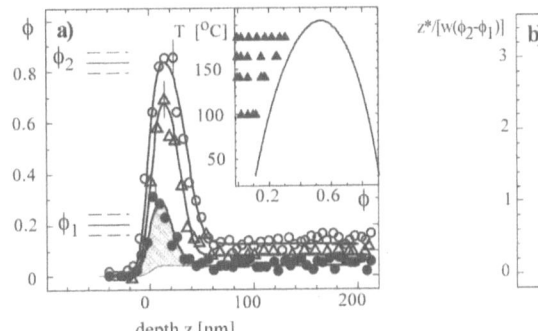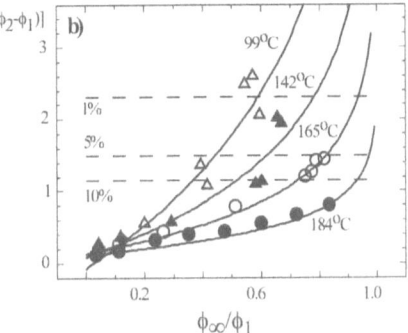

Fig. 19.a Local d66 volume fraction ϕ as a function of depth z, determined for the d66/h52 blend monolayers with different initial compositions following 2 h of annealing at 99 °C [16]. *Horizontal solid and dashed lines* indicate the respective binodal values ϕ_1 and ϕ_2 and their estimated uncertainty. The *hatched area* marks the d66 surface excess z*. The *inset* marks on the phase diagram bulk compositions ϕ_∞ for which z* was determined. *Solid curve* in the inset denotes binodal determined previously [91] and described by $\chi=(0.327/T+3.48\times10^{-4})(1+0.222\phi)$. **b** Segregation isotherm data [16] plotted as normalized surface excess $z^*/[w(\phi_2-\phi_1)]$ vs normalized bulk volume fraction ϕ_∞/ϕ_1. *Solid lines* are generated by Eq. 44 to fit the data. *Dashed horizontal lines* are normalized surface excess values for the bulk ϕ_1 phase enriched at the surface to the compositions ϕ_s such that $(\phi_2-\phi_s)/(\phi_2-\phi_1)$=10%, 5%, and 1%, respectively

segregation. This class of mixtures is characterized below by the results obtained for the d66/h52 blend. Surface excess z* has been determined as a function of bulk concentration ϕ_∞ at four different temperatures; the corresponding ϕ_∞ values are denoted as triangles (▲) in the d66/h52 phase diagram (inset to Fig. 19a). Characteristic profiles $\phi(z)$ at different bulk concentrations obtained at temperature T=99 °C (i.e., over 100 °C below the critical point T_c=204 °C) are shown in Fig. 19a. The shapes of these profiles are revealing. Although they are a convolution with the instrumental depth resolution (a Gaussian of ca. 8 nm HWHM at the depth z=0 nm), we may still observe that the surface concentration of the enriched layer ϕ_s attains the upper coexistence value ϕ_2 as the bulk concentration ϕ_∞ approaches the lower binodal value ϕ_1. The observed behavior indicates a complete wetting regime, where there are no energetic barriers to build up a macroscopic layer of the ϕ_2 phase, fully separating the bulk ϕ_1 phase from the surface.

The excess $z^*_1=z^*(\phi_\infty\to\phi_1)$, obtained for ϕ_∞ equal in the limit to ϕ_1, should diverge to infinity for the semi-infinite blend in the complete wetting regime. However, due to the finite thickness of our films z^*_1 would have a finite value, which depends on the overall amount of the material present in the film. In addition, a partial wetting regime would also be characterized by a finite z^*_1. These two wetting cases are certainly distinguishable within the Cahn construction analysis. It might be argued, however, that such distinction within Cahn approach is obtained at costs of assumptions inherent in it. We may get rid of most of these assumptions in a very simple method [16] where we compare the experimentally obtained surface excess limit z^*_1 with that expected $z^*_1(\phi_s)$ for the surface enriched profile, described by the coexistence profile (Eq. 11) between the phases ϕ_1 (in the bulk) and ϕ_2, but cut off by the surface at concentration ϕ_s:

$$z^*_1(\phi_s) = \int_{z(\phi_s)}^{\infty} \{\frac{1}{2}[(\phi_2 + \phi_1) - (\phi_2 - \phi_1)\tanh(z/w)] - \phi_1\}dz$$

$$= w(\phi_2 - \phi_1) \int_{x(\phi_s)}^{\infty} \frac{1 - \tanh x}{2}dx \tag{43}$$

where $x = z/w$. Here $z(\phi_s)$ is the position where the coexistence profile has a local concentration equal to ϕ_s. In principle this is just the free surface position $z=0$, but the formulation of Eq. (43) allows a more general discussion. Surface excess values z^*_1, corresponding to ϕ_s much lower than the upper coexistence value ϕ_2, would indicate partial wetting. The other limit, where z^*_1 corresponds to ϕ_s close to ϕ_2, would indicate the advent of the complete wetting regime.

To examine, using Eq. (43), the segregation isotherms $z^*(\phi_\infty)$ obtained at different temperatures [16] we re-plotted them using normalized scales, which completely adsorb the temperature-dependent interfacial width w and coexistence concentrations ϕ_1 and ϕ_2. Figure 19b shows how the normalized excess $z^*/[w(\phi_2-\phi_1)]$ varies with the normalized bulk concentration (ϕ_∞/ϕ_1). Dashed lines marked in this figure are the normalized excess values $z^*_1/[w(\phi_2-\phi_1)]$ obtained with Eq. (43) for the normalized difference between ϕ_2 and ϕ_s given by $[(\phi_2-\phi_s)/(\phi_2-\phi_1)]$, equal to 10, 5, and 1%, respectively. The experimental data at the lower temperatures overshoot these limits, suggesting that complete wetting regime has been attained even for the lowest temperature of 99 °C. This extends the previous observations [117] of complete wetting occurring at 150 °C, concluded by tracing the growth of the wetting layer from coexisting concentration ϕ_1. The data also seem to follow the expression for the normalized surface excess:

$$\frac{z^*}{w(\phi_2 - \phi_1)} \propto A' - \ln(1 - \frac{\phi_\infty}{\phi_1}) \tag{44}$$

which generates solid lines in Fig. 19b. This formula originates from the expression given by Cahn [153] for the thickness l of the surface enriched layer at temperatures above wetting transition point T_w. Cahn predicted a logarithmic divergence of the thickness l for bulk composition ϕ_∞ close to ϕ_1. Both the original Cahn formula and the approximation $l=z^*/(\phi_2-\phi_1)$ used to get Eq. (44) are valid only close to the complete wetting limit $\phi_\infty \to \phi_1$.

Finally, the Cahn construction analysis (see Sect. 3.1.1) performed for all segregation isotherms, yielded the "bare" surface energy derivative $(-df_s/d\phi)_s$ evaluated at different surface concentrations ϕ_s for all four studied temperatures (data sets corresponding to different temperatures are marked by different sym-

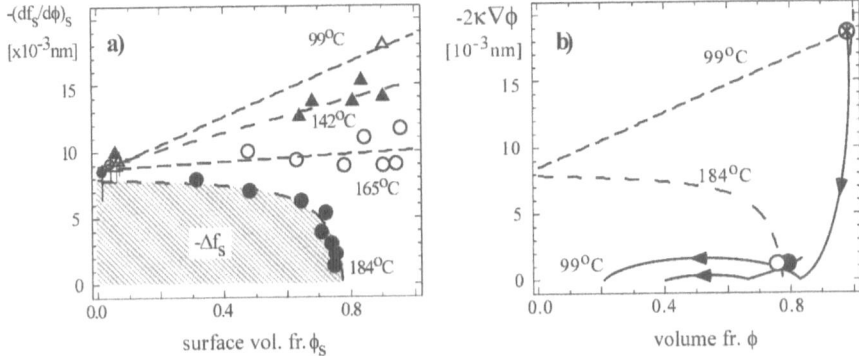

Fig. 20.a Results of the Cahn construction performed for the segregation data [16] of Fig. 19. Composition derivatives of "bare" surface free energy $(-df_s/d\phi)_s$ calculated for different temperatures (symbols: $\triangle$, $\blacktriangle$, $\bigcirc$, and $\bullet$ for T=99, 142, 165, and 184 °C, respectively) are fitted well by *dashed lines*, generated by the function $(\mu_1'+g'\phi_s)/(1+Y\phi_s)$. The *hatched area* marks the surface energy difference $-\Delta f_s$. **b** Surface energy derivatives $(-df_s/d\phi)_s$ (*dashed lines*) and trajectories $-2\kappa\nabla\phi$ (*solid lines*) plotted for T=99 °C and 184 °C. For T= 184 °C the surface boundary condition (Eq. 26) is met at point $\bullet$ at $\phi_s>\phi_2$, indicating complete wetting regime. If $(-df_s/d\phi)_s$ was *independent of temperature* (and equal to that found at 184 °C) then the boundary condition ($\bigcirc$) at 99 °C would correspond to partial wetting $(\phi_s<\phi_2)$. In practice, however, $(-df_s/d\phi)_s$ varies with temperature and the real boundary condition at 99 °C ($\otimes$) indicates complete wetting again

bols in Fig. 20a). It is evident that this relation is temperature dependent with a marked curvature at the highest temperature. For all studied olefinic blends a homographic function $(\mu_1'+g'\phi_s)/(1+Y\phi_s)$ (see dashed lines in Fig. 20a) has been found to describe calculated $(-df_s/d\phi)_s$ loci well. This functional form was related [120, 167] to the theoretical models [183–186] advocating the surface concentration gradients being present in surface free energy f_s. For all temperatures the $(-df_s/d\phi)_s$ loci start at the same value μ_1' at zero ϕ_s. They diverge, however, at higher ϕ_s where their slope g' decreases and the curvature Y increases monotonically with temperature. The existing theories [159, 177, 178, 194, 196], pointing out the temperature dependence of surface free energy f_s, cannot account for the observed changes in the shape of the $(-df_s/d\phi)_s$ vs ϕ_s relation. They [177, 178] may however explain qualitatively the observed variation of area under the $(-df_s/d\phi)_s$ plot, being the measure of the surface energy difference $|\Delta f_s|$ between blend components. $|\Delta f_s|$ increases at lower temperatures in qualitative accord with a simple lattice model [177] yielding quadratic f_s form: for instance, for (temperature independent) Q>1, Eq. (39) leads to larger μ_1 and hence larger $|\Delta f_s|$ at lower T (i.e., higher χ).

The final implication of the temperature variation of $(-df_s/d\phi)_s$ concerns the nature of the surface phase diagram and the location of the wetting transition point T_w. For further discussion we use the Cahn construction drawn in Fig. 20b for the highest (T=184 °C) and lowest (T=99 °C) studied temperature. While the concentration variations of the surface energy derivative $(-df_s/d\phi)_s$ (dashed curves) are just re-plotted from Fig. 20a, the trajectories $-2\kappa\nabla\phi$ vs ϕ (solid curves) are calculated for bulk compositions kept at lower coexistence values $\phi_\infty=\phi_1$ corresponding, similar to interaction parameter χ, to both temperatures considered. The intersection point (denoted as ●) of the $(-df_s/d\phi)_s$ and $-2\kappa\nabla\phi$ relations at T=184 °C corresponds to a complete wetting situation with surface concentration $\phi_s>\phi_2$. At lower temperature (T=99 °C) the solid curve representing the trajectory $-2\kappa\nabla\phi$ vs ϕ is shifted so that the positions of binodal concentrations ϕ_1 and ϕ_2, are moved out on the concentration axis to lower and higher values, respectively, while the extremum hump between them is bigger. *If the $(-df_s/d\phi)_s$ relation was independent of temperature* and equal to that determined for T=184 °C then its intersection with the trajectory $-2\kappa\nabla\phi$ (marked as ○) would correspond to a partial wetting. At intermediate temperatures a second order wetting transition would be observed. As seen in practice, however, the $(-df_s/d\phi)_s$ relation is not temperature independent but moves up at lower temperatures. As a result, the intersection point (denoted as ⊗) of $(-df_s/d\phi)_s$ and $-2\kappa\nabla\phi$ relations at T=99 °C corresponds again to the complete wetting situation. Thus the temperature dependence of the "bare" surface free energy parameters would *prevent* a complete-to-partial wetting transition from occurring. The extrapolation of this effect to lower temperatures suggests that complete wetting in the d66/h52 (and also d86/h75) mixture should occur even at room temperature.

A very similar complete wetting behavior characterizes [16, 145] the segregation isotherm data obtained for the free surface of the d86/h75 mixture even 75 °C below its critical point T_c. Again, the segregation data have been obtained basing on equilibrium profiles $\phi(z)$ corresponding to the one-phase region of the phase diagram. Earlier, a partial wetting behavior had been suggested [117] for the same blend to explain a phase inversion observed in thin bilayers of two coexisting phases of this mixture, with a less favored phase initially in contact with the free surface. These earlier dynamic studies [117] may be re-interpreted in terms of the free surface completely wetted by the d86 rich phase and the substrate interface with similar preference of both phases.

3.1.2.3.2
Blends With a Small Surface Energy Difference Δf_s

At least three of the studied blends [16, 145]: h86 ($N_{86}=1520$)/d75 ($N_{75}=1625$, deuteration extent e=0.4), d75/h66 ($N_{66}=2030$) and h66/d52 ($N_{52}=1510$ as well as e=0.34) may be described by a rather small Δf_s driving surface segregation. In an extreme case of the lowest Δf_s magnitude the enrichment-depletion effect is expected, as observed for the h66/d52 blend (see the next section). Here we characterize this class of mixtures with the results obtained for the h86/d75 blend. Surface excess z* has been determined [145] as a function of bulk concentration ϕ_∞ at two different temperatures; the corresponding ϕ_∞ values are denoted as open circles (○) in the h86/d75 phase diagram (see inset to Fig. 21a). Surface segregation of the h86 component with a local concentration $\phi(z)$ has been stud-

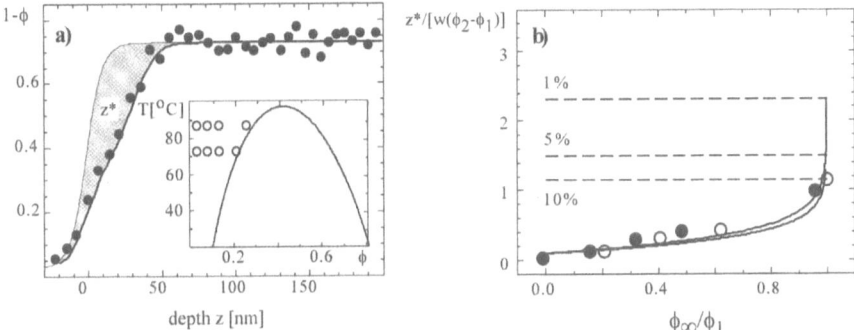

Fig. 21.a Local d75 volume fraction $(1-\phi)$ as a function of depth z (● points and the *thick solid line*), determined for the h86/d75 blend monolayer with 30%h86 following 46 h of annealing at 86 °C [145]. The *thin solid line* corresponds to the non-annealed sample with identical bulk composition. The *hatched area* marks the h86 surface excess z*. The *inset* marks the h86 bulk compositions ϕ_∞ for which z* was determined. *Solid curve* in the inset denotes binodal, determined previously [91] and described by $\chi=(0.547/T-8\times10^{-5})(1-0.217\phi)$. **b** Segregation isotherm data (corresponding to T=72 °C (○) and T=86 °C (●)) plotted as normalized surface excess $z^*/[w(\phi_2-\phi_1)]$ vs normalized bulk volume fraction ϕ_∞/ϕ_1. *Solid lines* are generated by Eq. 44 to fit the data. *Dashed horizontal lines* are normalized surface excess values for the bulk ϕ_1 phase enriched at the surface to the compositions ϕ_s such that $(\phi_2-\phi_s)/(\phi_2-\phi_1)=10\%$, 5%, and 1%, respectively

ied using the depth profiling method tracing the profile $(1-\phi(z))$ of the deuterated blend constituent d75. As usual, the surface excess z* is determined (hatched area in Fig. 21a) as the difference between the real profile of the annealed sample and the profile corresponding to a non-annealed sample kept at the same bulk composition. This reference profile also marks the real locus of the free surface on the depth scale. Obtained segregation isotherm data are plotted in Fig. 21b to show how normalized excess $z^*/[w(\phi_2-\phi_1)]$ varies with normalized bulk concentration (ϕ_∞/ϕ_1). We see at once that only at $\phi_\infty/\phi_1\to1$ does the normalized experimental excess $z^*/[w(\phi_2-\phi_1)]$ reach the $z^*/[w(\phi_2-\phi_1)]$ limit corresponding to surface concentration $\phi_s=\phi_1+0.9(\phi_2-\phi_1)$. This is in contrast to the results obtained for blends with strong force driving the surface segregation (see Fig. 19b).

The Cahn analysis, performed for both segregation isotherms, yielded the "bare" surface energy derivatives $(-df_s/d\phi)_s$ evaluated at different surface concentrations ϕ_s for T=72 and 86 °C (marked by ○ and ● points in Fig. 22a), respectively. The calculated $(-df_s/d\phi)_s$ loci are again well described by a homographic function which generated the dashed lines in Fig. 22. For both temperatures the corresponding trajectories $-2\kappa\nabla\phi$ vs ϕ (solid curves in Fig. 22b) are calculated for bulk compositions kept at lower coexistence values $\phi_\infty=\phi_1$. When the temperature is reduced, the trajectory $-2\kappa\nabla\phi$ shifts so that the binodal values ϕ_1 and ϕ_2 move out to lower and higher values, respectively, while the extremum hump between them becomes bigger. This variation is accompanied by such a change in the $(-df_s/d\phi)_s$ vs ϕ_s relation, which *cannot* now *prevent* the sit-

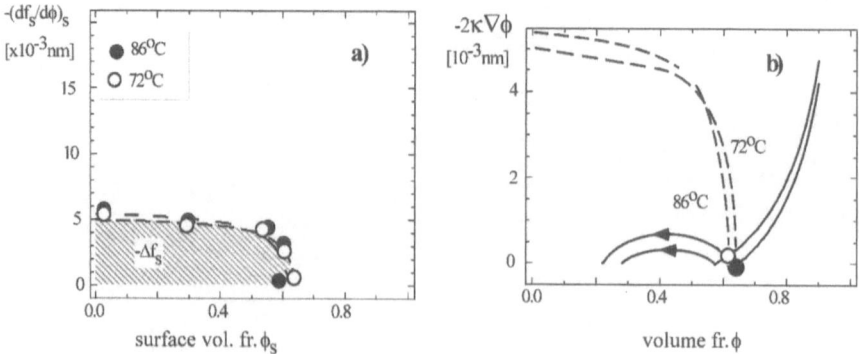

Fig. 22.a Results of the Cahn construction performed for the segregation data [145] of Fig. 21. Composition derivatives of "bare" surface free energy $(-df_s/d\phi)_s$ calculated for two temperatures (symbols: $\bigcirc$, and $\bullet$ for T=72 °C and 86 °C, respectively) are fitted by *dashed lines* generated by the function $(\mu_1'+g'\phi_s)/(1+Y\phi_s)$. The *hatched area* marks the surface energy difference $-\Delta f_s$. **b** Surface energy derivatives $(-df_s/d\phi)_s$ (*dashed lines*) and trajectories $-2\kappa\nabla\phi$ (*solid lines*) plotted for T=72 °C and 86 °C. For T=86 °C the surface boundary condition (Eq. 26) is met at point $\bigcirc$ at $\phi_s>\phi_2$ indicating complete wetting regime. In turn the boundary condition ($\bullet$) at 72 °C corresponds to $\phi_s\approx\phi_2$ and to a wetting transition point T_w!

uation characteristic for the wetting transition: the intersection point of $(-df_s/d\phi)_s$ and $-2\kappa\nabla\phi$ relations is located at T=86 °C ($\bigcirc$ point in Fig. 22b) for surface concentration $\phi_s>\phi_2$ *but* at T=72 °C ($\bullet$ point in Fig. 22b) for $\phi_s\approx\phi_2$!

This Cahn construction (Fig. 22) suggests that for the h86/d75 blend the wetting point T_w ($\approx$72 °C) may be located close to the critical point T_c (=97 °C). Similar conclusions are suggested by the Cahn analysis made for the h66/d52 and d75/h66 blends. Our very recent experiments [215], studying the growth of the surface excess layer from coexistence composition, seem to confirm these findings.

3.1.2.3.3
Temperature Variation of Δf_s

Existing mean field theories [177, 178] may only qualitatively describe the temperature variation of the surface energy Δf_s (surface tension $\Delta\gamma$, see Eq. 34) difference between blend components. On the other hand it is well known that the surface tension γ of simple liquids and polymers varies with temperature [216] according to an empirical formula found by Guggenheim [217]:

$$\gamma(T)=\gamma_0(1-T/T_{cr})^{11/9} \tag{45}$$

where T_{cr} is an imaginary critical temperature of the polymer. A theoretical justification for the expression at Eq. (45) has been found by Cahn and Hilliard, who studied the surface free energy of pure liquid in equilibrium with its vapor [53]. Parameters γ_0=53.7 mJ/m^2 and T_{cr}=1030 K were determined earlier [216]

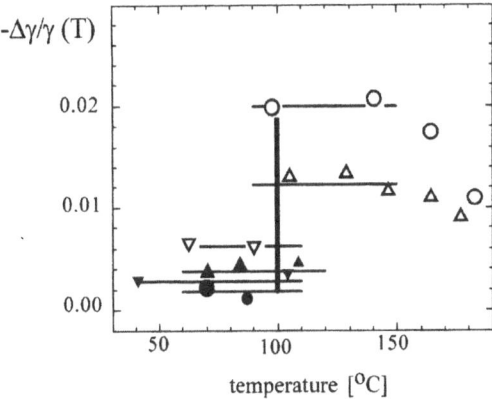

Fig. 23. The relative surface tension difference $-\Delta\gamma/\gamma$(T) between components of six poly-olefinic blends grouped in three microstructurally identical pairs x_1/x_2 ($x_1 > x_2$): 66/52 (O, ●), 86/75 (△, ▲) and 75/66 (▽, ▼). *Open and solid symbols* correspond to blends with deu-terated more (x_1) and less (x_2) branched components, respectively. *Large and small symbols* correspond to previously determined whole segregation isotherms and singular surface ex-cess data, respectively [16, 120, 145]. γ(T) is given by Eq. (45) for polyethylene. *Solid lines* denote average values for each blend at T_{ref}=100 °C (*thick bar*)

for the linear polyethylene (E), which corresponds (for x=0) to the random co-polymers $E_{1-x}EE_x$ discussed here.

In Fig. 23 we present the temperature variation of a relative difference in the surface tension $\Delta\gamma/\gamma$ (T) between components of six different blends. For each temperature, $\Delta\gamma$ was calculated on the basis of the corresponding Cahn con-struction and integrated curve $(-df_s/d\phi)_s$ vs ϕ_s (Eq. 34). For γ(T) we take the re-lation for polyethylene (x=0), since no data on γ (T) for any $E_{1-x}EE_x$ are available. This cumulative plot suggests that the surface tension difference $\Delta\gamma$ monotoni-cally decreases with increasing temperature. In addition, the rate of the reduc-tion in $\Delta\gamma$ seems to be well described by the formula at Eq. (45) in the wide tem-perature range between 40 °C and 150 °C. Apparently, the parameter T_{cr} is not very sensitive to the change of copolymer composition x. The values of the ratio $\Delta\gamma/\gamma$ (T), obtained at a reference temperature T_{ref}=100 °C, constitute the basis of the more general characterization of the surface energy parameter χ_s presented in Sect. 3.1.2.5.

3.1.2.4
Surface Enrichment-Depletion Duality

A conventional understanding of the surface segregation from polymer blends is that the surface should be enriched in the component with lower "bare" sur-face free energy f_s, regardless of the value of bulk composition ϕ_∞. This is how-ever true only when $(-df_s/d\phi)_s$ does not change its sign when surface concentra-tion is varied (see Fig. 14b). For such blends, surface enrichment in the same

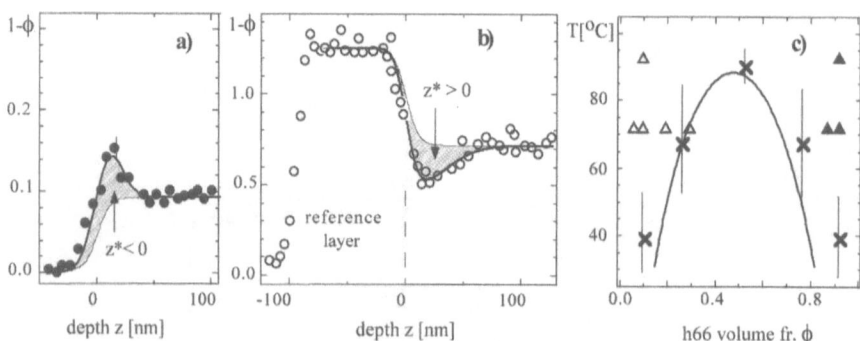

Fig. 24.a–b Typical d52 volume fraction (1–φ) vs depth z profiles indicating: **a** a *depletion*; **b** an *enrichment* in the h66 component, obtained for 90%h66/10%d52 and 30%h66/70%d52 monolayers annealed at 71 °C for 16 and 43 h, respectively [175]. *Hatched areas* mark *positive* (b) and *negative* (a) values of the h66 surface excess z*. The free surface locus (z=0) is yielded by: – the profile itself (a); – a profile of the control layer measured prior to the annealed sample (as in Fig. 21a); – the interface created by a reference layer positioned on top of the annealed sample (b); **c** a phase diagram as outlined by previously determined coexistence compositions [91] (*solid line* described by $\chi=(0.452/T-1.2\times10^{-4})(1+0.031\phi)$) and coexistence temperatures [138] (X points). Bulk compositions in one phase region are marked where the surface *enrichment* (Δ symbols) or *depletion* (▲ points) in h66 is concluded

component is expected for both sides of the miscibility gap ($T<T_c$), i.e., for the bulk concentrations $\phi_\infty\leq\phi_1$ and $\phi_\infty\geq\phi_2$. This enrichment is, however, smaller for $\phi_\infty\geq\phi_2$ in comparison to $\phi_\infty\leq\phi_1$ case, due to the asymmetry in relevant trajectories $-2\kappa\nabla\phi(\phi)$ (compare, e.g., solutions for $\phi_\infty=\phi_1$ and $\phi_\infty=\phi_2$ in Fig. 14). This is why, in practice, the segregation isotherms $z^*(\phi_\infty)$ have been determined [16, 92, 120, 165–167, 170, 173, 174] only for a limited ($\phi_\infty\leq\phi_1\leq\phi_c$) bulk concentration range.

In recent work Jerry and Dutta [176] analyzed within a mean field approach, conditions for critical wetting to occur. They have concluded that second order wetting transition has to be accompanied by the surface enrichment-depletion duality at least for the assumed quadratic form (Eq. 28) of the "bare" surface energy f_s. Polymer mixture is expected to exhibit surface enrichment when the bulk composition ϕ_∞ is below certain value Q and depletion when it is above Q. Both critical wetting and duality originate in their model from the surface energy derivative changing its sign at surface concentration $0<Q<1$, as was explained in Sect. 3.1.1 (see Fig. 15 for $Q=Q_e=Q_d$). Similar magnitude of the surface enrichment and the depletion is expected for anti-symmetric situations on related Cahn constructions (Fig. 15b,d). This best experimental situation is achieved both for Q close to critical composition ϕ_c and for the surface energy derivative $(-df_s/d\phi)_s$ having the form of an odd function $F(\phi_s-Q)=-F(Q-\phi_s)$. The second condition may be equivalently expressed by the requirement of very small surface energy difference Δf_s.

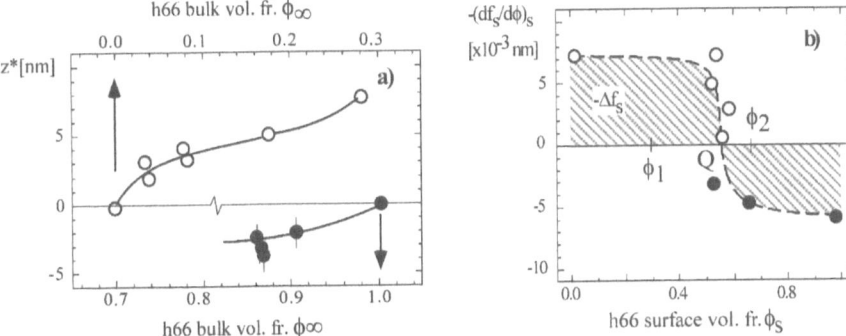

Fig. 25.a Determined at T=71 °C *positive* (○ points) and *negative* (● symbols) values of the h66 surface excess z* as the function of the h66 bulk composition ϕ_∞. *Solid lines* are back-calculated on the basis of the determined function $(-df_s/d\phi)_s$ vs ϕ_s. **b** Results of the Cahn construction performed for the segregation data [175] of a. Composition derivatives of "bare" surface free energy $(-df_s/d\phi)_s$ calculated for *positive* and *negative* z* values (symbols: ○ and ●, respectively) are well described by a single function equal to zero at Q=0.56. The *hatched areas* mark the positive and negative contribution to the surface energy difference $-\Delta f_s$. Since $\phi_1 < Q < \phi_2$, the situation is similar to that of Fig. 15 with $Q_e = Q_d$

We have observed the enrichment-depletion duality in the h66/d52 blend at the two temperatures of 71 °C and 92 °C. Figure 24 presents typical equilibrium profiles (1-ϕ) of the d52 component composing the minority and the majority of the mixture. While a surface enrichment in d52 is directly seen in the first case (Fig. 24a), a depletion in d52 in the second case is always observed while using a reference deuterated layer, pointing at the exact surface (z=0) locus. Such a reference layer is measured prior to the annealed sample (as for the h86/d75 blend in Fig. 21a) or simultaneously with it as its top cover (Fig. 24b). The h66 surface excess z* has been determined as a function of h66 bulk concentration ϕ_∞ and plotted for T=71 °C in Fig. 25a with two branches describing the enrichment (z*>0) and depletion (z*<0), respectively. Figure 24c locates on the phase diagram the bulk concentrations ϕ_∞ at which each of the z* signs is observed (△ for z*>0 and ▲ for z*<0).

The Cahn construction analysis, performed for the T=71 °C isotherm, yielded the "bare" surface energy derivatives $(-df_s/d\phi)_s$ evaluated at different surface concentrations ϕ_s (marked in Fig. 25b by ○ and ● points for the h66 enrichment and depletion, respectively). The calculated loci are well described by a single function, going to zero at surface volume fraction Q=0.56. The situation on the Cahn construction is topologically equivalent to that of Fig. 15, which was discussed above, although Q differs slightly from ϕ_c=0.47 and the $(-df_s/d\phi)_s$ vs ϕ_s relation is not exactly anti-symmetric yielding, after integration, a nonzero ($-\Delta f_s$) value.

The shape of the obtained "bare" surface energy derivative $(-df_s/d\phi)_s$ recalls similar curves obtained for other blends (see Figs. 20 and 22), which are however

equal to zero at concentrations Q more distinctly different from the corresponding critical compositions ϕ_c. Test experiments performed for the h86/d75 blend,
with $Q \approx 0.62$ and $\phi_c = 0.42$ most closely resembling the situation in h66/d52, have
revealed no detectable signs of the surface being enriched in the d75 component.
They suggest that the negative "depletion" branch of the $(-df_s/d\phi)_s$ vs ϕ_s relation
on the Cahn construction for h86/d75 blend (Fig. 22a) which, although possible
in principle, is much weaker that similar branch determined for the h66/d52
mixture (Fig. 25b).

Similar to other blends composed of random olefinic copolymers $E_{1-x}EE_x$,
here the more branched (more flexible) component h66 (i.e., with higher x) also
has lower "bare" surface energy f_s than its more linear counterpart d52 ($\Delta f_s < 0$).
However, in contrast with other isomeric blends studied so far, here the more
linear polyolefine may also be enriched at the free surface in the case when it
constitutes the minority of the mixture.

The enrichment-depletion duality occurs, when the surface energy derivative
$(-df_s/d\phi)_s$ changes its sign at the concentration Q, such that $0<Q<1$. This is possible, e.g., for segregation driven entirely by the missing neighbor effect in
blends composed of polymers with similar cohesive energies $\varepsilon_{AA} = \varepsilon_{BB} \neq \varepsilon_{AB}$ [178,
179]. For the h66/d52 mixture we have however $\varepsilon_{AA} \neq \varepsilon_{BB}$ and $\varepsilon_{AB} = (\varepsilon_{AA} \varepsilon_{BB})^{1/2}$
(leading to $Q>1$ for purely enthalpic segregation to the free surface (see Eq. 36)).
Hence the observation of the duality effect implies another mechanism driving
the segregation in addition to the missing neighbor effect. This is in accord with
conclusion drawn in the next section.

3.1.2.5
Effect of Deuterium Substitution on Surface Segregation

The "staining" of individual molecules, obtained by a replacement of hydrogen
for deuterium (used to create a contrast required in experiments with condensed matter) may lead to drastic changes in the phase behavior of the studied
systems. It has been widely recognized that this effect influences bulk interactions in polymer mixtures, as is discussed in Sect. 2.2.3. Here we describe first
experimental results [145] on the role of deuterium labeling on surface interactions in blends which are not isotopic mixtures.

We have studied binary blends dx_1/hx_2 of random olefinic copolymers $x \equiv (E_{1-x}
EE_x)_N$, with one blend constituent protonated (hx) and the other deuterated
(dx). The blends examined were grouped in four pairs of structurally identical
mixtures x_1/x_2 but with a swapped isotope labeled component (dx_1/hx_2 and
hx_1/dx_2). For such blend pairs the bulk interaction parameter χ (and hence also
the critical point T_c) has been found (see Sect. 2.2.3 and references therein) to be
higher when the more branched (say $x_1>x_2$) component is deuterated, i.e.,
$\chi(dx_1/hx_2)>\chi(hx_1/dx_2)$ or $T_c(dx_1/hx_2)>T_c(hx_1/dx_2)$ (see Fig. 9). An identical pattern is exhibited here by the force driving the segregation at the free surface. This
is illustrated in Fig. 26a,b where the composition vs depth profiles of the more
branched (x_1) component are shown for blend pairs with swapped isotope

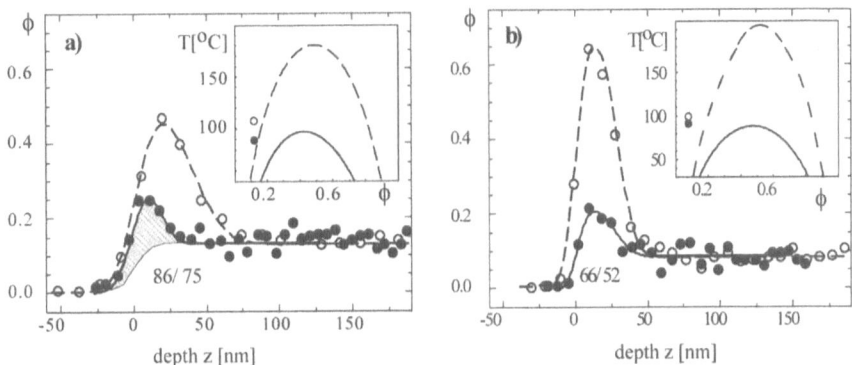

Fig. 26.a Composition ϕ vs depth z profiles of x=86% copolymer through annealed to equilibrium films of d86/h75 mixture ($\bigcirc$ points and a *dashed line*, T=106 °C, [16]) and of h86/d75 blend ($\bullet$ points and a *solid line*, T=86 °C, [145]). The surface excess z* (*shaded area* marked for h86/d75) increases when h86/d75 mixture is exchanged for d86/h75 blend in line with the shift of phase diagram presented in the *inset* (*solid and dashed lines* for h86/d75 and d86/h75, respectively). In the *inset* the symbols: $\bigcirc$ and $\bullet$ mark the bulk compositions of profiles presented for d86/h75 and h86/d75 blends, respectively. **b** Composition ϕ vs depth z profiles of x=66% copolymer through annealed to equilibrium films of d66/h52 ($\bigcirc$ points and a *dashed line*, T=99 °C, [16]) mixture and h66/d52 ($\bullet$ points and a *solid line*, T=89 °C, [175]) blend. The surface peak (and the excess z*) increases when h66/d52 mixture is exchanged for d66/h52 blend in line with the shift of phase diagram presented in the *inset* (*solid and dashed lines* for h66/d52 and d66/h52, respectively). In the *inset* the symbols: $\bigcirc$ and $\bullet$ mark the bulk compositions of profiles presented for d66/h52 and h66/d52 blends, respectively

"stained" component. The surface x_1 excess z* is evidently larger for the blend with deuterated more branched component (see profiles marked by $\bigcirc$ and fitted by dashed lines) than for the mixture with labeled more linear constituent (see profiles marked by $\bullet$ and fitted by solid lines). Figure 26 also suggests that the magnitude of changes in z*, caused by the exchange of the deuterium labeled component, seems to be larger for 66/52 than for 86/75 blends. Similarly the magnitude of changes in T_c (see insets to Fig. 26) is larger for 66/52 than for 86/75 mixtures. The question of correlation between isotope swapping effects in bulk and surface interactions is addressed more quantitatively below.

Our further analysis is based on the relative difference of the surface tension $-\Delta\gamma/\gamma$ between blend components, evaluated at a reference temperature $T_{ref}=$ 100 °C and presented in Fig. 23. Determined $-\Delta\gamma/\gamma$ values (see Fig. 23) are always higher for the blends with deuterated more branched components (denoted by open symbols) than for their counterparts with isotope labeled more linear constituents (denoted by solid symbols). This pattern is almost undetectable for the 52/38 blend pair with the $-\Delta\gamma/\gamma$ values at T_{ref} equal to 1.31(19)% and 1.23(24)% for d52/h38 and h52/d38 blends, respectively. The change in $-\Delta\gamma/\gamma(T_{ref})$, caused by isotopic swapping, is the highest for the 66/52 blend pair (symbols: $\bigcirc$ and $\bullet$ in Fig. 23) and decreases through 86/75 (symbols: $\triangle$ and $\blacktriangle$) and 75/66 (symbols:

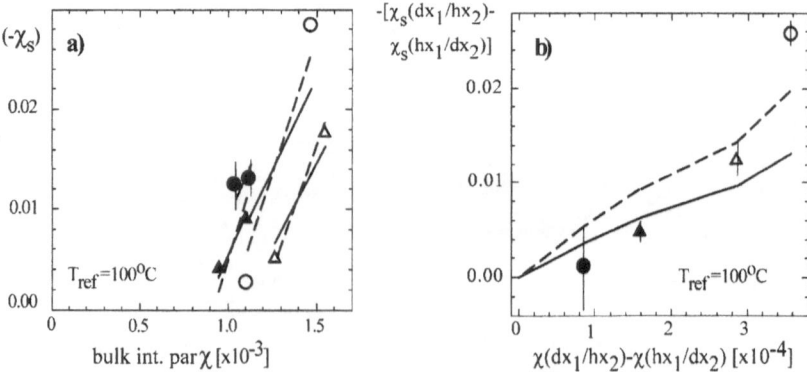

Fig. 27.a Surface energy difference parameter χ_s plotted as a function of corresponding bulk interaction parameter χ, both at $T_{ref}=100$ °C, for four pairs of microstructurally identical blends x_1/x_2 ($x_1 > x_2$): 66/52 (○), 86/75 (△), 75/66 (▲) and 52/38 (●). For each pair a point with higher ($-\chi_s$) and χ values corresponds to the mixture with deuterated more branched component (dx_1/hx_2). **b** The change in surface $-[\chi_s(dx_1/hx_2)-\chi_s(hx_1/dx_2)]$ and bulk parameter $[\chi(dx_1/hx_2)-\chi(hx_1/dx_2)]$ due to the swap of the isotope labeled component for four blend pairs x_1/x_2: 66/52 (○), 86/75 (△), 75/66 (▲) and 52/38 (●). The *solid and dashed lines* mark prediction of the model described in the text, which includes the missing neighbor effect for $z_s/(z_b-2)=0.5$ and 0.75, respectively

∇ and ▼) to the 52/38 blend pair. The same sequence gives a monotonic decrease in the difference of critical temperature $T_c(dx_1/hx_2) - T_c(hx_1/dx_2)$ (yielding values equal to 116, 84, 68, and 27 °C, respectively).

A more detailed insight into this relation might be obtained when comparing surface χ_s and bulk χ interaction parameters both evaluated for all the examined blend pairs at the same reference temperature $T_{ref}=100$ °C (see Fig. 27). The χ_s values are just recalculated (with Eqs. 45, 34, and 35) from $-\Delta\gamma/\gamma(T_{ref})$ values described above. The χ values are basing on critical point data: $\chi(T_{ref})=\chi_c T_c/T_{ref}$ (see Sect. 2.2.3). This is based on the assumption that only enthalpic contribution (Eq. 16) is relevant for bulk interactions.

Figure 27a presents surface parameter $(-\chi_s)$ plotted vs bulk parameter χ for four blend pairs (dx_1/hx_2, hx_1/dx_2) (different symbols correspond to different blend pairs). In turn, Fig. 27b compares the change in surface and bulk parameters caused by the exchange of the blend component "stained" by deuterium. While Fig. 27b reveals that the change in χ_s upon isotope swapping seems to be related with similar change in χ, Fig. 27a shows that some relations between absolute values of χ_s and χ may exist only within each blend pair x_1/x_2.

These facts may be explained as follows. Both the measured absolute values of bulk interaction parameter χ and their changes due to isotope swapping are well described by enthalpy-based arguments alone (see Sect. 2.2.3). The involved solubility parameter formalism describes practically *whole* [145] effective *bulk* interaction parameter χ (Eq. 16a). It yields, however, only *one* (enthalpic) *con-*

tribution to the *surface* interaction parameter $\chi_s^{H,mn}$ due to the missing neighbors effect (Eq. 37):

$$\chi = \frac{V}{k_B T}[\delta_A - \delta_B]^2 \quad , \quad \chi_s^{H,mn} = \frac{z_s}{z_b - 2}\frac{V}{k_B T}[\delta_A^2 - \delta_B^2] \tag{46}$$

Another, entropy related, contribution to surface interaction parameter χ_s^S is advocated by current theories applicable to olefinic blends (see Fig. 18 and the related text). The essential parameter for most of these theories, considering configurational and packing entropy effects, is the statistical segment length a, eventually combined with the segmental volume V (see Sect. 3.1.2.2). The change of the isotopic status of olefinic molecules seems to hardly affect [72, 218] their statistical segment length and changes only to a small extent the segmental volume ($\Delta V/V < 0.3\%$ for structurally similar polybutadiene [68]). Thus, it may be argued that such entropic forces driving to segregation would be insensitive, at least with a good approximation, to the deuterium labeling of polymer macromolecules:

$$\chi_s = \chi_s^{H,mn} + \chi_s^S, \qquad \chi_s^S(dx_1/hx_2) \approx \chi_s^S(hx_1/dx_2) \tag{47}$$

As a result, the changes in surface parameter χ_s due to isotope swapping should be described by enthalpy-based arguments alone:

$$\chi_s(dx_1/hx_2) - \chi_s(hx_1/dx_2) \approx \chi_s^{H,mn}(dx_1/hx_2) - \chi_s^{H,mn}(hx_1/dx_2) \tag{48}$$

Each blend pair x_1/x_2 has a characteristic entropic contribution χ_s^S hardly sensitive to the isotope swapping. Within each such pair the larger bulk interaction parameter χ corresponds to a bigger difference between solubility parameters δ_A and δ_B and, hence, to an increase in both the enthalpic contribution $\chi_s^{H,mn}$ and the surface parameter χ_s itself. This explains qualitatively the remarks made for Fig. 27a.

The change upon isotope swapping of surface parameter χ_s is represented by the change in $\chi_s^{H,mn}$. In turn, $\chi_s^{H,mn}$ is obviously closely related to the change in bulk parameter χ (see Eq. 46), as variations in solubility parameter δ_i are related to corresponding modifications of cohesive energy density δ_i^2. This clarifies the situation in Fig. 27b.

More quantitative comparison is also possible. On the basis of bulk interaction parameters χ and available data on PVT properties of pure blend components we have calculated the corresponding solubility parameters δ_i (see Fig. 10a and the related text). These allowed us to evaluate the corresponding enthalpic contributions $\chi_s^{H,mn}$ (see Eq. 46), the magnitudes of the swapping effect $\chi_s(dx_1/hx_2) - \chi_s(hx_1/dx_2)$ (Fig. 27b), and total surface parameters χ_s best fitting the experimental data (Fig. 27a). Their relations with the corresponding bulk quantities are marked in Figs. 27 by solid and dashed lines for the ratio of surface and bulk coordination number assumed as $z_s/(z_b-2)=0.5$ and 0.75, respectively. The change in the coordination number at the surface $z_s/(z_b-2)$ assumed

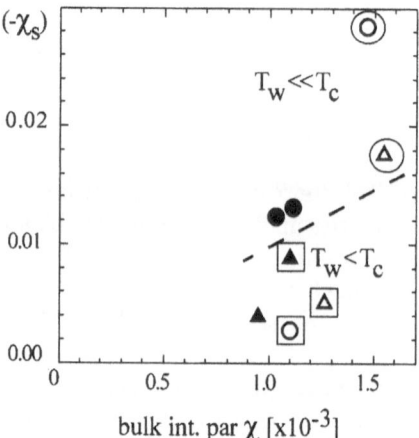

Fig. 28. The surface $-\chi_s$ vs bulk χ parameter re-plotted from Fig. 27 for four microstructurally identical blends x_1/x_2 ($x_1 > x_2$): 66/52 ($\bigcirc$), 86/75 ($\triangle$), 75/66 ($\blacktriangle$) and 52/38 ($\bullet$). For each pair a point with higher $(-\chi_s)$ and χ values corresponds to the mixture with deuterated more branched component (dx_1/hx_2). The Cahn constructions performed for available segregation isotherm data suggest two types of behavior: i) a relatively extended (even to ca. 100 °C) critical point wetting region ($T_w \ll T_c$) is observed for blends with $(-\chi_s, \chi)$ loci on the *plot surrounded by circles*; ii) wetting point located close to T_c ($T_w < T_c$) is suggested for blends with $(-\chi_s, \chi)$ loci on the *plot surrounded by squares*

here might be compared with the value 0.25 or 0.3 usually taken in simple lattice models [219]. Higher values (e.g., 0.5) are obtained in experiments [220] even for surface regions much thicker than one lattice site assumed in the model used here.

The performed calculations, yielding reasonable predictions for the behavior observed in Fig. 27a,b, also reveal the absolute magnitudes of both contributions to the surface energy difference parameter χ_s. For all the blends examined here, the average values – $\chi_s^{H,mn} = -0.092(3)$ for the enthalpic and $\chi_s^S = +0.080(4)$ for the entropic contributions – are obtained for the ratio $z_s/(z_b-2) = 0.5$. Thus, while the *enthalpic* force drives the more *flexible* molecules (with higher x) to the surface, its *entropic* counterpart, almost in magnitude, favors the *stiffer* (with lower x) chains at the surface. An equal weight for both segregation forces is in fact expected by a recent self consistent mean field analysis [214].

A swap of the blend component "stained" with deuterium results in the interrelated changes in surface χ_s and bulk χ interaction parameter. The model presented above accounts for this relation, using the solubility parameter formalism introduced in Sect. 2.2.3. Large magnitudes of the swapping effect for bulk and, hence, surface interactions observed in 66/52 and 86/75 blend pairs lead to the observed disparity in surface phase behavior noticed within each pair (see Fig. 28). Extended critical point wetting region is concluded for d66/h52 and d86/h75 mixtures with $(-\chi_s) > 0.014$, while its very drastic reduction is concluded for the blends with swapped isotope labeled component, i.e., for h66/d52 and h86/d75 mixtures with $(-\chi_s)<0.01$. Smaller magnitudes of the swapping effect might result in similar surface phase diagrams for both structurally identical mixtures x_1/x_2; for instance the wetting point located close to T_c is concluded for d75/h66 and h75/d66 mixtures (for the last mentioned blend only single z^* data are available), both with $(-\chi_s)<0.01$. Still the theoretical model, which may account for the entropic contribution χ_s^S and, hence, also for the absolute value of the surface energy parameter χ_s is missing. Only in a frame of such a model would we be able to predict, in a priori fashion, the surface phase diagram for the given blend components.

3.1.3
Summary and Conclusions

Equilibrium composition profiles $\phi(z)$, describing segregation at the surface of a binary homopolymer blend, are monitored by depth profiling techniques with a resolution ρ comparable with polymer coil dimensions. Mean field theory generates these profiles $\phi(z)$ similarly as was done previously for the phase coexistence in the bulk. Now the profiles $\phi(z)$ are, however, cut off at the surface by a boundary condition involving a short-ranged "bare" surface contribution f_s to free energy. Different assumptions, such as inclusion of long-range surface interactions or model extension beyond the long wavelength limit, do not improve fits with experimental data significantly, while much more complicated calculus is involved. A clear picture of segregation is obtained within so-called Cahn construction, topologically equivalent to the phase portrait relation $d\phi/dz$ vs ϕ. It allows us, based on available segregation isotherm data, to determine the compositional derivative of the "bare" surface energy $(-df_s/d\phi)_s$ and, further, to analyze a surface phase diagram. The results of the Cahn analysis are trustworthy only if coexistence conditions were previously determined for the studied blends (Sect. 3.1.2.1).

Integrated "bare" surface energy derivative $(-df_s/d\phi)_s$ yields three equivalent parameters: Δf_s, χ_s, and $\Delta\gamma$ which describe the surface tension difference between blend pure components expressed in units of length, energy/$k_B T$ (dimensionless) and energy/area. Segregation to a neutral surface, such as vacuum//blend interface, is expected (missing neighbor effect) and related with the difference in cohesive energies between blend components. While such enthalpic (and short-range) arguments describe well [197] χ_s obtained for a symmetric isotopic polystyrene blend [92] (Sect. 3.1.2.2), they fail to account for observations made for mixtures with chain length disparity or statistical segment length disparity. A large variety of entropic contributions to "bare" surface free energy f_s is postulated and related with effects due to polymer configurations being perturbed by the surface, polymer density gradient present at the surface, or different segment packing at the surface (Sect. 3.1.2.2).

Standard theoretical approaches, using the graphical Cahn construction [8, 153, 176, 221] or its numerical analogs [15, 162, 187] to analyze the surface phase diagram, have assumed that the parameters of the "bare" surface energy f_s are independent of temperature. We have performed [16, 145] the first detailed study on the role of temperature in the segregation. It was done for a few mixtures composed of olefinic random copolymers. They show (Sect. 3.1.2.3) that with decreasing temperature the difference Δf_s increases monotonically, with a rate comparable, at a wide temperature range, to that empirically established by Guggenheim [217] for surface tension $\gamma(T)$. The related change in $(-df_s/d\phi)_s$ has significant consequences for surface phase diagram modifying it as compared to the predictions yielded by models with a constant f_s. For blends with a large overall magnitude of Δf_s, its increase with lowered T prevents the wetting transition T_w from occurring even 100 °C below critical point T_c [16] (Sect. 3.1.2.3). On the contrary, for mixtures with relatively small magnitude of Δf_s, its increase with lowered T cannot prevent a situation typical for a partial wetting from occurring on the Cahn plot. Therefore T_w is concluded here to be located close to T_c (Sect. 3.1.2.3).

According to a conventional view the surface of the mixture is enriched in the component with lower surface energy, regardless of the value of the bulk composition. This is not necessary true as it is shown by our observation [175] of surface enrichment-depletion duality (Sect. 3.1.2.4). This first experimental evidence confirms previous Monte Carlo results [178, 179] and mean field predictions [176] that polymer mixture can exhibit surface enrichment when bulk composition is below a certain value Q, and depletion when it is above Q. Mean field theory finds this phenomenon to be related [176] to a second order wetting transition. Both the duality and the critical wetting would occur for $(-df_s/d\phi)_s$ changing its sign at concentration Q. As a result, positive and negative contributions to Δf_s almost cancel each other, yielding a very small Δf_s value.

It is widely recognized that isotope labeling of polymer macromolecules may also result in drastic changes in bulk thermodynamics of mixtures which are not isotopic blends. We have performed [145] first studies focused on similar effect in surface interactions of such systems (Sect. 3.1.2.5). A few olefinic mixtures with stiffness (i.e., statistical segment length) disparity have been examined. They are grouped in pairs of structurally identical blends with exchanged the component "stained" by deuterium. Observed, upon isotope swapping, change in determined surface parameter χ_s has the same pattern as the change in bulk parameter χ (Sect. 2.2.3), i.e., larger values are determined when the blend component with a lower cohesive energy value (and a smaller statistical segment length) is "stained" by deuterium. The magnitudes of the changes in χ_s and χ are related. A simple model accounting for these observations is proposed (Sect. 3.1.2.5). It suggests that both enthalpic and entropic contributions to χ_s are of comparable importance, but only the enthalpic term is sensitive to isotope swapping.

3.2
Mixtures Between Two Interfaces

So far we have described *separately* the segregation phenomena occurring at both surfaces of thin films composed of binary polymer mixtures. Now we discuss situations where a semi-infinite mixture approach cannot be applied. We present a brief summary of the established mean field theory and the experimental status quo.

3.2.1
Mean Field Theory

Consider a binary polymer mixture confined in a thin film of thickness D with both surfaces, left (L) and right (R), exerting specific short-ranged surface fields. Related "bare" surface contributions to free energy are denoted by $f_s^L(\phi_s^L)$ and $f_s^R(\phi_s^R)$. The overall free energy F (per site volume Ω and area A normal to the surface) is expressed on the analogy of Eq. (24) as [6, 60, 177, 219, 222]:

$$\frac{F}{k_B TA/\Omega} = f_s^L(\phi_s^L) + f_s^R(\phi_s^R) + \int_0^D dz \{\Delta F_M[\phi(z)]$$

$$-\Delta\mu\phi(z) + \kappa(\phi)[\nabla\phi(z)]^2\}$$

(49)

A closely related form has been used earlier [93, 221] within Landau theory.

A solution to this variational problem is given by a differential equation describing the profile $\phi(z)$ and two surface boundary conditions. The equation defining the trajectory $-2\kappa\nabla\phi$ vs ϕ is given on the analogy of Eq. (25):

$$-2\kappa(\phi)\frac{d\phi}{dz}(\phi) = \pm 2\sqrt{\kappa(\phi)\Delta f(\phi; \chi, \Delta\mu, \phi_b)} =$$

$$\pm 2\sqrt{\kappa(\phi)[\Delta F_M(\phi) - \Delta F_M(\phi_b) - \Delta\mu(\phi - \phi_b)]}$$

(50)

Here the parameter $\Delta f(\phi; \chi, \Delta\mu, \phi_b)$, defined by Eq. (8), describes the excess free energy needed to create a (local) unit volume of a blend with composition ϕ from a reservoir with a flat profile ($\nabla\phi(z)=0$) kept at composition ϕ_b. For symmetric profiles ϕ_b corresponds directly to the concentration in the middle of the thin film $\phi(z=D/2)$. The chemical potential difference $\Delta\mu$, bounded by the relation $\Delta\mu=\partial\Delta F_M/\partial\phi(\phi_\infty)$ for a semi-infinite mixture $\phi_b=\phi_\infty$, may be varied now.

The left (L) and right (R) boundary conditions are given by

$$-\frac{df_s^L}{d\phi}(\phi_s^L) = \mu_{1L} + g_L\phi_s^L = \pm 2\sqrt{\kappa(\phi_s^L)\Delta f(\phi_s^L; \chi, \Delta\mu, \phi_b)}$$

(51a)

and

$$\frac{df_s^R}{d\phi}(\phi_s^R) = -\mu_{1R} - g_L\phi_s^R = \pm 2\sqrt{\kappa(\phi_s^R)\Delta f(\phi_s^R; \chi, \Delta\mu, \phi_b)}$$

(51b)

respectively. We have expressed them in terms of simple, quadratic in composition, forms of "bare" surface free energy f_s^L and f_s^R (see Eq. 28).

Surface concentrations ϕ_s^L and ϕ_s^R, determined by conditions (51), as well as the trajectory $-2\kappa\nabla\phi$ (Eq. 50), specify the overall thickness D of the profile:

$$D = \int_{\phi_s^L}^{\phi_s^R} (-2\kappa)d\phi /(-2\kappa\nabla\phi) \tag{52}$$

and the average concentration $<\phi>$ along the profile:

$$<\phi> = \frac{1}{D}\int_{\phi_s^L}^{\phi_s^R} (-2\kappa\phi)d\phi /(-2\kappa\nabla\phi) \tag{53}$$

The solution of the concentration profile $\phi(z)$ should be specified for given: temperature T, film thickness D, and the average blend composition in this film $<\phi>$. The parameters T and $<\phi>$, important in experiments, might be translated [60] into interaction parameter χ and the chemical potential difference $\Delta\mu$, more convenient in calculations. Thus, for say D, χ, and $\Delta\mu$ known and kept constant, the profile $\phi(z)$ may be obtained (Eq. 50) by varying the "reservoir" concentration ϕ_b until the boundary conditions (Eq. 51) are met. If a few solutions exist, the relevant ones are those with minimal overall free energy F (Eq. 49). Such a "shooting" procedure was developed by Flebbe et al. [60]. A numerical method which starts from an arbitrary assumed profile and modifies its discretized form until conditions equivalent to Eqs. (50), (51) and (53) are met has also been proposed recently by Eggleton [222]. The solutions yielded by this technique may however correspond to metastable states. Concentration profiles in thin films were also evaluated by other theoretical treatments [93, 118, 177, 219, 221].

The overall free energy of Eq. (49) could be rewritten [60] for a symmetric blend (with $N_A=N_B=N$):

$$\frac{\sqrt{N}}{a}\frac{F}{k_BTA/\Omega} = \frac{\sqrt{N}}{a}f_s^L(\phi_s^L) + \frac{\sqrt{N}}{a}f_s^R(\phi_s^R)$$

$$+ \int_0^{D/(a\sqrt{N})} dz\,\{\Delta F_M'[\phi] - N\Delta\mu\phi + \frac{\kappa}{a^2}[\frac{d\phi}{dz}]^2\} \tag{54}$$

where the free energy of mixing is modified ($\Delta F_M'$) and now corresponds to simple liquids:

$$\Delta F_M' = \phi\ln\phi + (1-\phi)\ln(1-\phi) + 2\frac{\chi}{\chi_c}\phi(1-\phi) \tag{55}$$

The form of Eq. (54) allows us to have better insight into the problem: it reflects scaling properties of a mixture between two interfaces. The behavior of such a blend is best characterized by a set of *scaling* parameters defined by

Eq. (54). These are film thickness $D/(aN^{1/2})$, interaction parameter χ/χ_c, the chemical potential difference $N\Delta\mu$, and "bare" surface energies $N^{1/2}/a\ f_s^L$ and $N^{1/2}/a\ f_s^R$. They enable [60] an easy comparison of finite size effects observed in different systems. Sometimes two such parameters, $D/(aN^{1/2})$ and χ/χ_c, are combined into the film thickness-to-interfacial width ratio $D/w=3(\chi/\chi_c-1)^{1/2}D/(aN^{1/2})$.

3.2.2
Finite Size Effects

3.2.2.1
Size Effects in Surface Segregation

We consider now the segregation to the surface bounding a polymer mixture from its left side. The blend is prepared in two geometries, corresponding to: i) a semi-infinite reservoir, ii) a thin film with thickness D and zero boundary condition exerted by the right surface. In the first case the bulk concentration $\phi_b= \phi_\infty$ determines unequivocally the chemical potential difference $\Delta\mu= \partial\Delta F_M/\partial\phi(\phi_\infty)$. On a related Cahn construction (dashed line in Fig. 29a) the single trajectory $-2\kappa\nabla\phi$ is cut off by a left boundary condition (Eq. 51) at ϕ_s^L. The corresponding profile (dashed line in Fig. 29b) exhibits considerable surface excess,

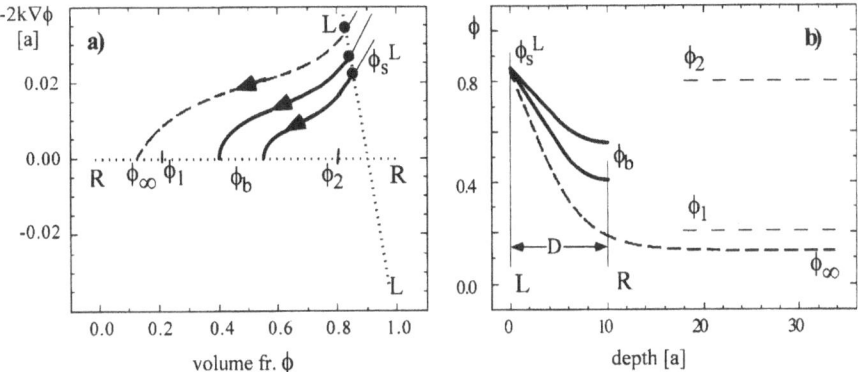

Fig. 29a,b. Size effect in surface segregation: **a** Cahn construction with trajectories $-2\kappa\nabla\phi$ vs ϕ; **b** corresponding profiles plotted at constant chemical potential difference $\Delta\mu=\mu_A-\mu_B$ (= 2.054×10^{-3}) for: a semi-infinite mixture (*dashed lines*) and for the same mixture [60] ($N_A= N_B=100$, $\chi=0.023$, $w=8.6a$) confined into a thin film with thickness D=10a (*pairs of solid lines*). Segregation occurs to the left surface of a semi-infinite reservoir and the left surface of a thin film. For the right surface of a thin film $f_s^R=0$ is assumed. The surface energy derivatives $(-df_s/d\phi)_s$ in (a) corresponding to the left (L) and to the right (R) surface are marked by *dotted lines*. The concentration, in the bulk, of a semi-infinite mixture ϕ_∞ is below the lower coexistence value ϕ_1, while two compositions of the flat profile's region in a thin film ϕ_b, calculated for the same $\Delta\mu$, are inside the bulk miscibility gap (ϕ_1,ϕ_2). The surface concentrations ϕ_s^L are comparable in both cases. The surface excess z^* is reduced considerably (see b) for a thin film as compared to the bulk characteristic for the assumed temperature region above the wetting point T_w. In the second case, the "shooting" method was used by Flebbe et al. [60], i.e., trajectories $-2\kappa\nabla\phi$ were tried out for constant $\Delta\mu$ (and equal to that one from the first case) but for varied concentration of the flat profile region at the right surface ϕ_b. Two stable equi-energetical solutions, which fulfill the condition at Eq. (51) for assumed D (here equal to 1.2 w), were obtained. They are represented by thick lines in Fig. 29. These two composition profiles (Fig. 29b) are revealing: first, the compositions ϕ_b are inside the miscibility gap corresponding to the mixture in the bulk – this change in coexistence conditions will be discussed later; second, although concentrations ϕ_b are above ϕ_∞, the surface excess values z^* for a finite film are visibly smaller than for a semi-infinite mixture. This originates mostly from the increase of ϕ_b as compared to ϕ_∞ while ϕ_s^L does not change too much. This tendency is preserved if other $\Delta\mu$–s are allowed leading to various values ϕ_b. Hence a decreasing excess $|z^*|$ with increasing composition ϕ_b is expected as observed in model calculations by Flebbe et al. [60] for a film thickness D=(1–3) w. Although a certain similarity with a semi-infinite case was sometimes concluded in other regions of segregation isotherm $z^*(\phi_b)$, as z^* was increased with ϕ_b, no behavior was observed which might properly indicate a partial or complete wetting regime.

Recently Monte Carlo simulations were performed [118] for the Ising lattice of a binary atomic (N=1) mixture at the complete wetting regime. They focused on the shape of segregation isotherms $z^*(\phi_b)$ as a function of the film thickness D. No logarithmic divergence of the segregation isotherm (with concentration approaching binodal $\phi_b \to \phi_1$) was observed for very thin films, contrary to thick layers regaining behavior typical for semi-infinite systems. Size effects were concluded to be negligible for the situation corresponding to films with thickness-to-correlation length ratio D/ξ=13 confined by two interfaces with a zero- and a non-zero surface field (corresponding to f_s^R=0 and $f_s^L \neq 0$). However, effects observed in simulations even for D/ξ=8.6 would hardly be detected by present profiling techniques with a finite precision in concentration determination (~0.01). Monte Carlo studies [118] also showed that for thin films the coexistence value ϕ_1, expected with increasing concentration ϕ_b, could be easily missed due to metastable states.

Singular experimental studies on the role of film thickness D on the surface segregation were performed by Hariharan et al. [172], who investigated deuterated polystyrene dPS segregating to the vacuum/ and silicon/blend interface of the isotopic mixture dPS/hPS (N≈4600). It was found that surface segregation was affected significantly if D was reduced below ca. four correlation lengths ξ of the system. Then surface excess decreased with decreasing film thickness D.

3.2.2.2
Size Effects in the Shape of "Intrinsic" Coexistence Profile

The coexistence curve of polymer blends is nowadays determined with a novel technique (see Sect. 2.2.1) tracing at different temperatures the composition

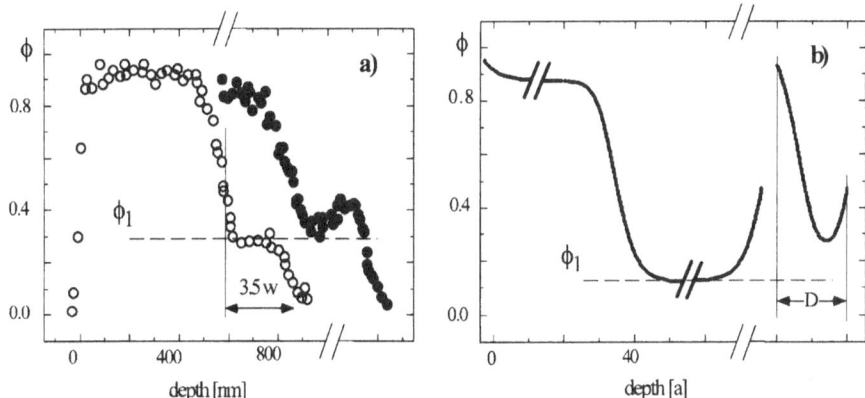

Fig. 30a,b. Size effect in the shape of coexistence profile $\phi(z)$: **a** the profiles [92] $\phi(z)$ of two identical bilayers composed of a thick dPS – and a thin (with a thickness of 3.5 w) hPS-film deposited on a gold substrate (○ points) and on SiO_2 backing (● points), following 27.7 days of annealing at T=170 °C. The isotopic mixture created by the used polymers is identical to that described by Figs. 4 and 7. The lower binodal value ϕ_1 is marked by a *dashed line*; **b** equilibrium concentration profiles $\phi(z)$ calculated [222] for a very thick (*on the left*) and a very thin (D=3.3w, *on the right*) layer, composed of symmetric blend ($N_A=N_B=100$ at $\chi=0.026$) with average composition $<\phi>=0.5$ bounded by asymmetric surfaces ($\mu_{1L}=0, g_L= -0.5a; \mu_{1R}=0.25a, g_R=-0.5a$)

profile $\phi(z)$ across a thin polymer film composed of two coexisting layers. Our studies [92], performed soon after this technique was established, indicated that such a coexistence profile may be modified by: i) the finite thickness of the layers used, ii) specific interactions with the confining interfaces. We illustrate our findings by two profiles $\phi(z)$ (Fig. 30a) determined for two identical asymmetric bilayers composed of a thick (>7 w) dPS and a thin (=3.5 w) hPS film. While the dPS film always faced the vacuum (where dPS is preferred), the hPS film covered the gold substrate in the first sample and a silicon oxide backing in the second one. Both samples were annealed for one month at 170 °C. Then the bilayer deposited on Au exhibits an hPS-rich region adjacent to the substrate with a lower coexistence value ϕ_1 determined by a visible plateau in the profile. Here, apparently, the hPS layer thickness is large enough so the lower coexistence composition ϕ_1 can be determined. The situation is, however, drastically changed when the zero surface boundary conditions, characteristic for gold substrate, are exchanged for those favoring strongly dPS at SiO_2. The segregation peak at the SiO_2 interface is visibly accompanied by a zero composition gradient appearing locally at concentration higher than ϕ_1! This precludes the determination of coexistence conditions corresponding to those in the bulk.

We were not able to calculate our observed profiles [92] in a priori fashion. Recently, Eggleton [222] performed the numerical calculations for a very similar model system with both asymmetric surfaces exerting fields (and related energy contributions $N^{1/2}/a\, f_s^{L,R}$) ten times stronger than reported in [92]. He varied the overall thickness of the bilayer D for constant average concentration $\langle\phi\rangle$, inter-action parameter χ, and "bare" surface energies f_s^L and f_s^R. Two of his profiles, corresponding to a very thick (D→∞) and thin (D=3.3 w) bilayer are presented in Fig. 30b. While the surface excess is visible for the thick bilayer, it cannot prevent the coexistence concentrations to be determined as long as large enough plateau regions are available. For films with D≤9.8 w zero composition gradient appears locally at concentration higher (by ~0.015) than ϕ_1, in qualitative agreement with our experimental data. Eggleton shows that his new numerical method [222] might be used to predict the variation of the coexistence profile in a priori fashion.

Recently, Binder et al. [118] considered the Ising lattice of a binary atomic (N=1) mixture confined in a very thin film by antisymmetric surfaces each attracting a different component. It was shown that the segregation of each blend component to opposite surfaces may create antisymmetric (with respect to the center of the film z=D/2) profiles $\phi(z)$ even for temperatures above critical point $T>T_c$, where flat profiles are expected when external interfaces are neglected. Such antisymmetric profiles would not be distinguished in experiments (with limited depth resolution) from coexisting profiles described by a hyperbolic tangent.

3.2.2.3
Size Effects in Coexistence Conditions

Different equilibrium situations are predicted for a polymer mixture confined in a thin film, depending on the fields exerted by its two bounding surfaces (see Fig. 5). Let us consider first a thin film with *two symmetric, neutral surfaces*. For a symmetric ($N_A=N_B=N$) mixture of polymers with the same contact energy between identical segments ($\varepsilon_{AA}=\varepsilon_{BB}\neq\varepsilon_{AB}$) Eqs. (39) and (40) lead to a specific form of the surface energy derivative $(-df_s^L/d\phi)_s=(-df_s^R/d\phi)_s=const\chi(1/2-\phi)$. Surface boundary conditions are executed for $(-df_s^L/d\phi)_s$ and for $(+df_s^R/d\phi)_s$. They are presented together with trajectory $-2\kappa\nabla\phi$ vs ϕ on the Cahn construction in Fig. 31a for $\Delta\mu=0$ and $\phi_b\geq\phi_1$, at temperature below T_c (i.e., $\chi>\chi_c$). Four partial trajectories solving the variational problem, originating at left and terminating at right boundary conditions, are visible on the plot. They are grouped into two pairs with equal profile thickness D (Eq. 52): (1–4, 3–2) and (1–2, 3–4). The first pair corresponds to antisymmetric profiles $\phi(z)$, which are considered to be metastable states for symmetric surfaces [60]. The second pair corresponds to two equi-energetical symmetric profiles $\phi(z)$ outlined in Fig. 31b. It is clear, that they exhibit surface *enrichment-depletion duality* (see Sect. 3.1.2.4), i.e., the minority component is always segregated to the surface. The average compositions $\langle\phi_1\rangle$ and $\langle\phi_2\rangle$ for coexisting profiles 1–2 and 3–4, respectively, are inside the miscibility gap (ϕ_1, ϕ_2) corresponding to the bulk. This effect is in some way due to the condition $\phi_b>\phi_1$ assumed on a Cahn plot in order to get finite D, but mainly due to the enrichment-depletion duality. For the overall average composition $\langle\phi\rangle$ in the thin film D such that $\langle\phi_1\rangle<<\langle\phi\rangle<<\langle\phi_2\rangle$ we expect

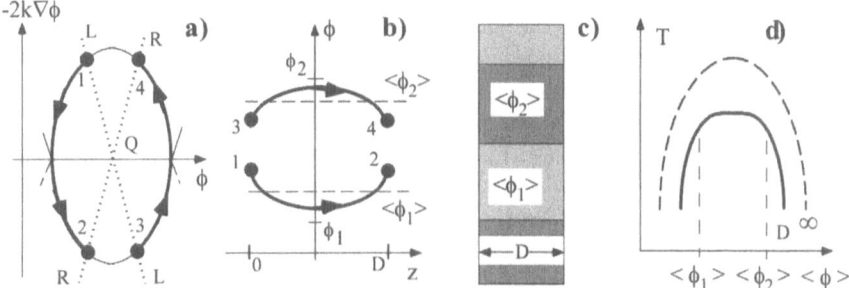

Fig. 31a–d. A thin blend film bounded by symmetric, neutral surfaces: **a** Cahn construction with trajectories $-2\kappa\nabla\phi$ (*solid lines*) plotted for $\Delta\mu=0$ and $\phi_b \geq \phi_1$. Surface free energy derivatives $(-df_s^L/d\phi)_s$ and $(+df_s^R/d\phi)_s$ due to missing neighbor effect at left (L) and right (R) surface are marked by *dotted lines*, which cross at Q=0.5. Boundary conditions at Eq. (51) are met at points 1,2,3,4. The variational problem is solved by two pairs of equi-energetical paths: a metastable (1–4, 3–2) and a *stable* (1–2, 3–4) one; **b** equi-energetical profiles 1–2 and 3–4, as determined by Cahn construction. The surface enrichment (for $\phi<Q$)-depletion (for $\phi>Q$) duality is visible. Duality plays an important role in shifting the average compositions $<\phi_1>$ for 1–2 and $<\phi_2>$ for 3–4 inside the bulk miscibility gap (ϕ_1, ϕ_2); **c** the equilibrium morphology with coexisting 2-dim domains: $<\phi_1>$ and $<\phi_2>$ with 1–2 and 3–4 profiles describing local concentration within each of them, respectively; **d** the corresponding phase diagram with average composition $<\phi>$ plotted vs temperature T for the bulk (∞) and for a thin film with thickness D

coexisting *two-dimensional domains*, $<\phi_1>$ and $<\phi_2>$, with interfaces separating them running across the film (see Figs. 31c and 5c). In Fig. 31d we plot the average composition $<\phi>$ vs temperature. The phase diagram defined in such a way for a thin film is symmetric around $\phi_c=1/2$ and inside the bulk miscibility gap.

Performed Monte Carlo simulations [178, 179] confirm the conclusion, that surface segregation of the minority component plays an important role in the suppression of the phase diagram. The segregation occurs on a scale of the correlation length ξ for any film with varied thickness D, but only for small D ($>4\xi$ [178]) it leads to visible divergence of the average compositions $<\phi_1>$ and $<\phi_2>$ from their bulk (binodal) values ϕ_1 and ϕ_2.

The picture presented above is not complete as it neglects non-mean field behavior of polymer blends in the temperature range close to T_c [149]. The Ising model predicts phase diagrams of thin films, which are more depressed and more flattened than those yielded by mean field approach (as marked in Fig. 31d). Both effects were shown by Monte Carlo simulations performed by Rouault et al. [150]. In principle, critical regions of phase diagrams cannot be described merely by a cross-over from a three- to two-dimensional (for very thin films) situation. In addition, a cross-over from mean field to Ising behavior should also be considered [6, 150].

Detailed (mean field) analytical calculations were performed by Tang et al. [219] to evaluate the shift in the critical temperature T_c^D of a thin film as a function of D. They considered a symmetric polymer blend confined by neutral walls, with "bare" surface energy parameters as given by Eq. (39) for Q=1/2. For long chains (N>100) and thin films (D<<0.2 Na) a linear relation between $|1/T_c^D-1/T_c|$ and 1/D was found. A 5 °C shift in T_c^D as compared to T_c=180 °C could be predicted from this work [219] for film composed of isotopic polystyrene blend (N=13,200) with the thickness D≈25 nm.

These evaluations are, however, not consistent with experimental data by Reich and Cohen [220] who have studied phase separation occurring in the mixture composed of low m.w. PS (N=104) and poly(vinylmethyl ether) PVME (N= 58) confined into thin films bounded by gold surfaces. Such data were taken [219] as representative for a blend between neutral surfaces. While the data follow the scaling $|1/T_c^D-1/T_c|$ vs 1/D, predicted by Tang et al., the experimentally determined slope of this relation was two orders of magnitude larger than the predicted one. Distinct shifts in phase diagram were found [220] for films thinner than 1 micron. It is possible that short range surface interactions assumed by Tang et al. might predict too small a magnitude of the shift in T_c^D.

For a symmetric blend confined into a thin film with two neutral surfaces (with surface fields related to a missing neighbor effect only) we obtain a phase diagram (see Fig. 31d) which is symmetric around ϕ_c=1/2. Average compositions of coexisting profiles correspond to the chemical potential difference $\Delta\mu$= 0, identical to that describing the coexistence curve in the bulk. A different situation is obtained when two *surfaces* bounding a thin film are *symmetric* but *selective*, i.e., both adsorb preferentially the same blend component. Here the coexisting profiles in a thin film occur for $\Delta\mu$ shifted from the zero bulk value. This shift in $\Delta\mu$ increases linearly for decreasing temperatures. This is the phenomenon of capillary condensation [60, 93]. The related phase diagram is displaced, as compared with bulk binodal, to lower temperatures and its critical point ϕ_c^D is moved from bulk ϕ_c=1/2 to the side rich in preferentially adsorbed polymer. Such behavior was observed for phase diagrams calculated within a mean field approach by Flebbe et al. [60]. These results predict, e.g., a 2 °C shift in T_c^D for a symmetric isotopic polystyrene blend (N=14,750) confined in a 500 nm thick film. However this prediction is valid only when both "external" interfaces have identical "bare" surface energies $N^{1/2}/af_s$ a few times stronger than experimentally observed at a free surface.

Very recent Monte Carlo simulations and self consistent mean field calculations [223] have shown that wetting properties might be reflected in the phase diagram of a blend confined between symmetric selective surfaces: Close to T_w a convex curvature is exhibited by the phase diagram on the side poor in preferentially adsorbed polymer. Also the temperature dependence of $\Delta\mu$ changes around the wetting point T_w.

3.2.2.4
Size Effects in Interfacial Width

The location of the wetting point T_w also has a direct significance for the phase diagram of a polymer blend confined between *antisymmetric surfaces* exerting

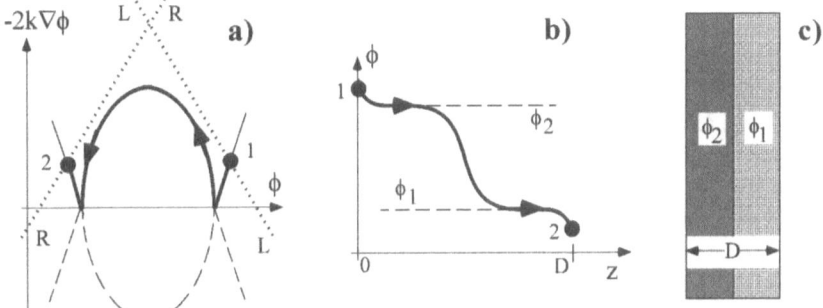

Fig. 32a–c. A thin blend film bounded by antisymmetric surfaces exerting opposing fields [93]: **a** Cahn construction with trajectories $-2\kappa\nabla\phi$ (*solid and dashed lines*) plotted for $\Delta\mu=$ 0 and (here) for $\phi_b\rightarrow\phi_1$. The "bare" surface free energy derivatives $(-df_s^L/d\phi)_s$ and $(+df_s^R/d\phi)_s$ due to left (L) and right (R) surface are marked by *dotted lines*. Boundary conditions (Eq. 51) are met at points 1 and 2; **b** the profile 1–2 as determined by Cahn construction (a) for a rather thick (due to the limit $\phi_b\rightarrow\phi_1$) film; **c** the corresponding bilayer equilibrium morphology with the interface between phases ϕ_1 and ϕ_2 running parallel to both surfaces

opposing fields, i.e., each attracting different blend component [93]. Here the critical point T_c^D is always located below T_w. At temperatures $T<T_c^D$, pairs of coexisting (stable) asymmetric profiles $\phi(z)$ appear for $\Delta\mu=0$, identical to $\Delta\mu$ yielded by bulk coexistence curve. Above T_c^D a single antisymmetric profile exists. At temperatures $T>T_w>T_c^D$ a single *"soft mode phase"* appears, characterized again by an antisymmetric profile $\phi(z)$ (plotted in Fig. 32 for a rather thick layer) but with anomalously large transverse correlation length $\xi_\parallel$ for concentration fluctuations in directions parallel to the substrate $\xi_\parallel=\xi_b\exp[D/(4\xi_b)]$, where ξ_b is the bulk correlation length at binodal $\xi_b=\xi(\phi_1)=\xi(\phi_2)$ (Eq. 6). The bilayer equilibrium structure allowed here (see Fig. 32c) is identical to the one observed (Fig. 5b) in many experiments studying the coexistence conditions.

As mentioned in Sect. 2.2.2, the effective interfacial width w_D characterizing the bilayer structure may be *broadened* beyond its "intrinsic" value w, yielded by a mean field theory (Eqs. 10 and 12). This is due to the capillary wave excitations causing the lateral fluctuation of the depth $I_e(x,y)$ corresponding to the midpoint of the "internal" interface between coexisting phases. This fluctuation is opposed by the forces due to "external" interfaces, which try to stabilize the position $I_e(x,y)$ in the center of the bilayer [6, 224, 225]. It was suggested recently [121] that the spectrum of capillary waves for a "soft mode phase" should be cut off by ξ_b and $\xi_\parallel$. This leads to the conclusion that the effective interfacial width w_D should depend on the film thickness D as: $(w_D/2)^2=\xi_b^2+\xi_b D/4$. Experimental data [121] obtained for olefinic blends (at T close to T_c) indeed show remarkable increase of the measured interfacial width from $w_D(D=160\text{ nm})=14.4(3)$ nm to $w_D=45(12)$ nm for thickness $D\approx660$ nm, where w_D levels off (because $\xi_\parallel$ is comparable with lateral sample dimensions). This trend is in qualitative agreement with the formula due to capillary oscillations in the "soft mode phase". However the mean field theory yields the "intrinsic" values of the coexistence tanh profile w=42 nm (Eqs. 8 and 10) and that of the bulk correlation length ξ_b=21 nm (Eq. 6), which are in fact ca. four times larger than those used (effectively) in [121] to fit the data. The exact mean field predictions describe well, without any additional contributions, the data for bilayers thicker than some 450 nm. Such films were used previously in experimental studies [91, 96] on coexistence conditions in other polyolefinic mixtures. The more direct size effect mechanism, relating the change in the slope of the antisymmetric ("intrinsic") profile with the shift of ϕ_b from the value ϕ_1 accomplished for Δ=0, is possible but effective only for very much reduced D (<about 6 w). The suggested scenario involving capillary wave excitations is the only so far known explanation for this effect observed for relatively thick (D<16 w!) layers of real polymer blends. This model is also in agreement with the results of Monte Carlo simulations [224]. The accordance could be improved for proper values of parameters such as lower cut off value of capillary wavelengths or the range of forces due to the "external" interfaces [6, 224, 225] (both parameters were related previously to ξ_b [121]).

Another type of dependence of effective interfacial width w_D on film thickness D was observed [130] for immiscible mixture of deuterated polystyrene (dPS) and poly(methyl methacrylate) (PMMA) (at T<<T_c): an increase, from w_D=1.8(4) nm for a dPS layer thickness D=6 nm to $w_D(D\approx100$ nm)=2.5(4) nm, follows the logarithmic dependence $w_D\propto\ln D$ (intrinsic interfacial width w= 1.5 nm). This may reflect [6, 224] long range forces acting from the "external" interfaces on the "internal" interface $I_e(x,y)$. On the contrary, the relation $w_D\propto D^{1/2}$ found for random olefines [121] corresponds [6, 224] to short range forces. We note also that capillary waves in dPS/PMMA system were observed [130] already for the thickness-to-intrinsic width ratio D/w<85!!!

3.2.3
Summary and Conclusions

A mean field theory has recently been developed to describe polymer blend confined in a thin film (Sect. 3.2.1). This theory includes both surface fields exerted by two "external" interfaces bounding thin film. A clear picture of this situation is obtained within a Cahn plot, topologically equivalent to the profile's phase portrait $d\phi/dz$ vs ϕ. It predicts two equilibrium morphologies for blends with separated coexisting phases: a bilayer structure for antisymmetric surfaces (each attracting different blend component, Fig. 32) and two-dimensional domains for symmetric surfaces (Fig. 31), both observed [94, 114, 115, 117] experimentally. Four finite size effects are predicted by the theory and observed in pioneer experiments [92, 121, 130, 172, 220] (see Sect. 3.2.2) focused on: (i) surface segregation; (ii) the shape of an "intrinsic" bilayer profile; (iii) coexistence conditions; (iv) interfacial width. The size effects (i)–(iii) are closely related, while (i) and (ii) are expected to occur for film thickness D smaller than 6–10 times the value of the "intrinsic" (mean field) interfacial width w. This "cross-over" D/w ratio is an approximate evaluation, as the exact value depends strongly on the strength and the range of surface fields exerted by both "external" interfaces (e.g., long-range forces are neglected in theoretical works reviewed here). The value D/w<6–10 is in accord with a majority ([92, 172], Fig. 6) but not all [220] of the rare experimental works. On the contrary the effective interfacial width between coexisting phases is predicted to be broadened monotonically with D beyond its "intrinsic" value w, even in macroscopically thick samples (with lateral dimensions large enough). Such a trend is mimicked quantitatively by the data from [130] and qualitatively by results of [121].

4
Diblock Copolymers Block-Anchored to Homopolymer Interfaces

4.1
Mean Field Theory

Diblock copolymers A-N immersed in a homopolymer P matrix segregate to its interfaces. One of the copolymer blocks ("anchor" moiety A) selectively attaches to the interface while the other ("buoy" block N) dangles out to form a brush like layer, providing a simple means for the realization of *polymer brushes* (see Fig. 33).

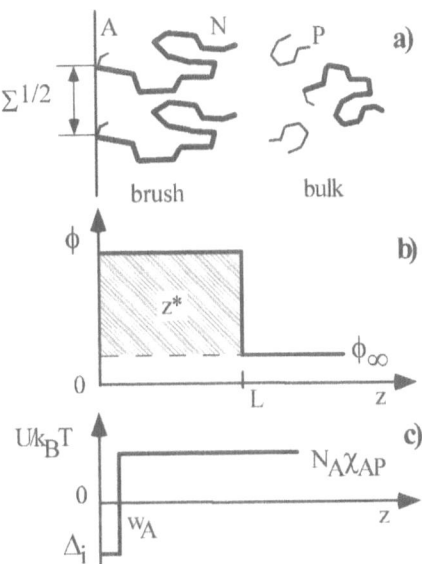

Fig. 33.a Schematic illustration of N-mer brush layer created by diblock copolymers A-N attached selectively to the interface by their "anchor" moiety A. Copolymers in the brush layer are in equilibrium with free diblocks incorporated in the bulk region of the sample abundant in homopolymer P. **b** The form of the diblock volume fraction vs depth $\phi(z)$ profile used in theoretical model. **c** Potential U(z) affecting the "anchor" moiety A and driving the diblock segregation

This name covers all polymer chains (diblocks and others) attached by one end (or end-block) at ("external") solid/liquid, liquid/air or ("internal") liquid/liquid interfaces [226–228]. Usually this is achieved by the modified chain end, which adsorbs to the surface or is chemically bound to it. Double brushes may be also formed, e.g., by the copolymers A-N, when the joints of two blocks are located at a liquid/liquid interface and each of the blocks is immersed in different liquid. A number of theoretical models have dealt specifically with the case of brush layers immersed in polymer melts (and in solutions of homopolymers). These models include scaling approaches [229, 230], simple Flory-type mean field models [230–233], theories solving self-consistent mean field (SCMF) equations analytically [234, 235] or numerically [236–238]. Also first computer simulations have recently been reported for brushes immersed in a melt [239].

Here we outline a mean field Flory-type model introduced by de Gennes [230] and developed by Leibler [231] and Aubouy and Raphaël [232]. This approach is less detailed than SCMF models but it captures the main features of the physics of segregated copolymers. Even though it makes a number of assumptions, which are a simplification in comparison with the SCMF models, its predictions of the main features (such as, e.g., variation of mean brush height L vs size and surface density σ of the diblocks) agree [226] well with those of more detailed SCMF calculations [236–238]. Because of clearness and simplicity it has been used as a basic framework for many experimental papers on brush conformation [240–245] and segregation properties of end-adsorbing polymers [246–255].

The segregation of diblock copolymers A-N to the interface (Fig. 33a) is driven by the spatial dependence of thermodynamic potential U [256], which affects N_A segments of the "anchor" block A (Fig. 33c). Unfavorable segmental interactions χ_{AP} of the "anchor" moieties with the homopolymer P matrix result in a relatively large and positive contribution $N_A\chi_{AP}$ to the potential U which is effective in the bulk of a sample. The "anchor" positioned at the interface, where it replaces matrix molecules P, may change interactions at the interface and reduce the interfacial energy. Therefore the gain Δ_i in the free energy, expressed as a negative contribution to U, and affecting diblocks at the interface is expected.

After the segregation process [235, 242, 245, 257, 258] is completed, the copolymers in the brush layer are in equilibrium with the free diblocks incorporated (at bulk concentration ϕ_∞) in the bulk region of the sample abundant in homopolymer P (see Fig. 33). Within the simple Flory-type picture the volume fraction Φ of brush ("buoy") N-mer chains is constant across the whole brush layer of thickness L:

$$\Phi = \frac{N\Omega}{L\Sigma} = \frac{Na}{L}(\frac{\Omega}{a^3})\sigma, \quad \Omega = a^3 \text{ or } \Omega = V. \tag{56}$$

Here Σ is the mean area per chain comprising the brush layer and $\Sigma^{1/2}$ is the mean inter-anchor spacing (see Fig. 33). Ω is the lattice site volume approximated by the cube of statistical segment length a^3 [230, 231] or, more precisely, by the reference segmental volume V of involved polymers [246, 248]. The surface coverage σ by the brush N-mer chains is usually defined as an areal density σ= a^2/Σ.

Following de Gennes and Leibler, we write down the free energy G of one chain in an interface-attached brush:

$$\frac{G}{k_B T} = \frac{L\Sigma}{\Omega} [\frac{1}{P}(1-\Phi)\ln(1-\Phi) + \chi\Phi(1-\Phi)] + \ln(\frac{N\Omega}{\Sigma a}) + \frac{3}{2}\frac{L^2}{Na^2} + \Delta_i. \qquad (57)$$

The first term is the free energy of mixing (compare with Eq. 1) between the brush N-mer and the homopolymer P, evaluated for all lattice sites of the volume $L\Sigma$ comprising one brush chain. The entropic contribution $\Phi\ln\Phi/N$ is missing since the translational freedom of the anchored N chain is lost [230]. The first term of Eq. (57) accounts for the combinatorial entropy of P-mer present in the brush layer as well as for the interaction between matrix P and brush N segments, which is specified by the parameter χ. The next term is also entropic in origin and it is associated with the two-dimensional translational freedom of the N-mer chains localized at the interface [231]. The third term represents the increase in the elastic energy accompanying the chains stretching beyond their unperturbed dimensions. All brush chains are assumed to be stretched at the same distance L from the surface. Their unperturbed configurations are characterized by the mean square end-to-end distance $R_0(N)=aN^{1/2}$ or, related to it, radius of gyration $R_g(N)=R_0/6^{1/2}$ – both typical for ideal N chains immersed in polymer P melt, provided that $P>N^{1/2}$ [230] (a more precise criterion $P>N^{1/2}(1+2\chi N^{1/2})^{-1}$ is given by Aubouy and Raphaël [232]). The last term Δ_i characterizes the free energy change associated with the presence of the "anchor" block at the interface.

4.1.1
Brush Conformation L(σ)

At low values of the surface coverage σ, each anchored N-mer chain is essentially independent of its neighbors. It forms a *mushroom* (see Fig. 34, and region "I" in Fig. 35), the size of which is characterized by R_0. When mushrooms start to overlap, a continuous brush layer is created. This happens for the inter-anchor spacing $\Sigma^{1/2}\cong R_0$ and the overlap density $\sigma_1\cong N^{-1}$. To consider the brush conformation in this regime, we re-write Eq. (57) in the limit of small volume fraction Φ of N-mer chains [232]:

$$\frac{G}{k_B T} = -\frac{N}{P}(1-P\chi) + \frac{La^2}{\Omega\sigma}[\frac{1-2P\chi}{2P}\Phi^2 + \frac{1}{6P}\Phi^3] + \frac{3}{2}\frac{L^2}{Na^2} + \ln(N\frac{\Omega}{a^3}\sigma) + \Delta_i \quad (58)$$

While the first one and the last two terms do not depend on the brush height L, all the other terms would govern the brush conformation. The second term is so-called osmotic contribution corresponding to two body (Φ^2) and three body

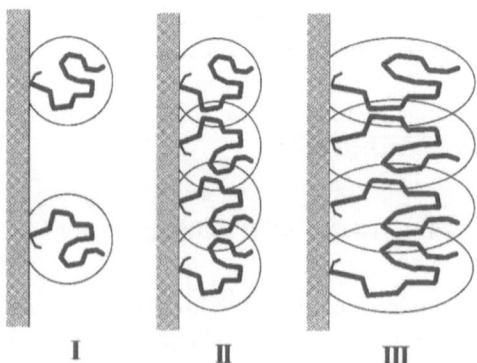

Fig. 34. End-anchored N-mers forming: "mushrooms" (I) and continuous, unswollen (II) and stretched (III), brush layer

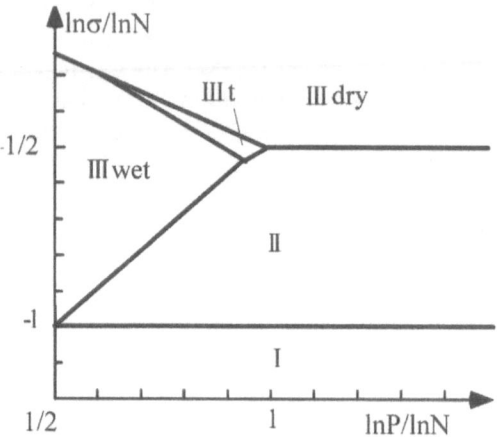

Fig. 35. Schematic conformation diagram (lnσ/lnN vs lnP/lnN) for N-mer brush exposed to P-mer chains, structurally identical but with a different isotopic status (χ-dependent details of the diagram calculated for the data from [245], describe well the situation also for other experimental reports). The brush height L is described as the power law $L \propto \sigma^q$, with the exponent q characterizing different regimes: q=0 (I and II), 1/3 (III wet), 1/2 (III t), 1 (III dry). The cross over between different regions occurs for coverage values $\sigma_1 \cong N^{-1}$ (I/II), $\sigma_2 \cong PN^{-3/2}/(1-2P\chi)$ (II/III wet), $\sigma_3 \cong (1-2P\chi)^2 P^{-1/2}$ (III wet/III t), $\sigma_4 \cong P^{-1/2}$ (III t/III dry), $\sigma_5 \cong N^{-1/2}$ (II/III dry)

(Φ^3) repulsive interactions between N-mer segments. These interactions, characterized by the virial coefficients [4, 230, 232] $\Omega(1-2P\chi)/P$ (excluded volume parameter) and Ω^2/P, are the driving forces for the swelling of the brush. They are opposed by the elastic energy contribution (the third term in Eq. 58).

Let us consider a case, when two body interactions are dominant in the osmotic contribution. The onset of brush stretching is expected for the amplitude of the osmotic term being comparable with the one of the elastic contribution (with $L \cong R_0$) [230, 232]. This would happen for the surface coverage $\sigma_2 \cong PN^{-3/2}/(1-2P\chi)$. We note that the threshold value σ_2 for the onset of stretching depends on the homopolymer degree of polymerization P. This is the consequence of the excluded volume interactions being "screened" by homopolymer P chains, as expressed by the virial coefficient. While for the brushes immersed in a good solvent (corresponding to very small P) chain stretching occurs already when separate mushrooms start to overlap, in melts ($P > N^{1/2}$) the chain stretching is not observed until the surface coverage σ_2 is reached (see Fig. 35). The "screened" brush regime, where anchored chains overlap but the brush stretching is still halted (i.e., $L \cong N a^{1/2}$), is marked as "*II*" in Fig. 35.

Minimizing, with respect to the brush height L, the relevant brush chain free energy G (Eq. 58) with three body interactions neglected leads to

$$L(\sigma) = 6^{-1/3} N \Omega^{1/3} \sigma^{1/3} [\frac{1-2P\chi}{P}]^{1/3} \qquad \text{(``III wet'')} \qquad (59)$$

This prediction characterizes the *wet brush* regime ("*III wet*" region in Fig. 35). Brush layer is accessible here to host matrix P molecules. When $1 >> |2P\chi|$, then the brush swelling is *entropic* in nature and brush thickness is given by $L \cong N\Omega^{1/3}(\sigma/P)^{1/3}$. *Enthalpy*-driven brush swelling with $L \cong N\Omega^{1/3}\sigma^{1/3}(-2\chi)^{1/3}$ is expected for a favorable enthalpy of mixing between matrix P and brush N chains ($-2P\chi >> 1$) [240]. The corresponding free energy G for the wet brush is obtained using Eqs. (58) and (59):

$$\frac{G}{k_B T}\Big|_{wet} = -\frac{N}{P}(1-P\chi) + \frac{3^{4/3}}{2^{5/3}} N (\frac{\Omega}{a^3})^{2/3} [\frac{1-2P\chi}{P}]^{2/3} \sigma^{2/3} + \ln(N\frac{\Omega}{a^3}\sigma) + \Delta_i \quad (60)$$

For big enough areal density σ (and $P < N$) three body interactions become the dominant term in the osmotic contribution to Eq. (58). Minimizing Eq. (58), with two body interactions neglected, yields the brush thickness $L(\sigma)$:

$$L(\sigma) = 3^{-1/2} Na(\frac{\Omega}{a^3})^{1/2} \sigma^{1/2} P^{-1/4} \qquad \text{(``III t'')} \qquad (61)$$

typical for the wet brush regime with *small excluded volume parameter* $\Omega(1-2P\chi)/P$ (region "*III t*" in Fig. 35). The cross over between regions "III wet" and "III t" (described by Eqs. 59 and 61, respectively) occurs for $\sigma_3 \cong (1-2P\chi)^2/P^{1/2}$.

Finally, for very high coverage σ, the N-mer chains stretch to maintain the constant density of the brush layer. In this *dry brush* regime (marked as "*III dry*" in Fig. 35) the brush volume fraction $\Phi = 1$ and from Eq. (56) stems the brush thickness L:

$$L(\sigma) = Na(\frac{\Omega}{a^3})\sigma \qquad \text{(``III dry'')} \qquad (62)$$

The corresponding free energy per brush chain has the form of Eq. (57), neglecting the first (free energy of mixing) term:

$$\left.\frac{G}{k_B T}\right|_{dry} = \ln(N\frac{\Omega}{a^3}\sigma) + \frac{3}{2}N(\frac{\Omega}{a^3})^2\sigma^2 + \Delta_i . \tag{63}$$

The cross over between regimes "III dry" and "III t" (described by Eqs. 62 and 61, respectively) is expected for $\sigma_4 \cong 1/P^{1/2}$. The "III t" regime, characteristic for wet brush and dominating three body interactions, exists in very narrow and usually neglected coverage range $\sigma_3 < \sigma < \sigma_4$ (see Fig. 35). The dry brush regime might be reached directly from the "screened brush" case, for very high degree of polymerization of the matrix homopolymer $P > N$ and at the coverage $\sigma_5 \cong 1/N^{1/2}$ (see Fig. 35 for the transition from region "II" to "III dry").

4.1.2
Segregation Isotherm $\phi_\infty(\sigma)$

The basic assumption of the de Gennes-Leibler approach is that the copolymers segregated at the interface are in equilibrium with the free diblocks incorporated (at concentration ϕ_∞) in the host matrix homopolymer P in the bulk of the sample (see Fig. 33). It means that the chemical potentials μ_{brush} and μ_{bulk} of the copolymers at the interface and in the bulk are equal, i.e.:

$$\mu_{brush}(\sigma) = \mu_{bulk}(\phi_\infty) . \tag{64}$$

The brush chemical potential (per chain) μ_{brush} is expressed by [231]:

$$\mu_{brush} = G + \sigma\frac{\partial G}{\partial\sigma} . \tag{65}$$

Note that for various brush regimes different forms for the free energy G (see, e.g., Eqs. 60 and 63) and for the brush chemical potential μ_{brush} are obtained.

The full relation for the chemical potential of the copolymers in the bulk μ_{brush} can be obtained from the Flory-Huggins energy of mixing between diblock copolymers A-N and homopolymers P [259, 260], when interaction parameters: χ_{AP}, χ_{AN}, and χ ($\chi := \chi_{NP}$) are specified. In most experiments brush N-mers and homopolymer P-mers are microstructurally identical and differ only in the isotopic status. Related isotopic interaction parameter χ is usually much smaller than parameters χ_{AP} and χ_{AN}. Assuming [254] $\chi_{AN} = \chi_{AP} + \chi$ and neglecting volume fraction of the "anchor" moieties in the bulk, the expression for μ_{bulk} is obtained in the form

$$\frac{\mu_{bulk}}{k_B T} = \ln\phi_\infty + 1 - N_c(\frac{1-\phi_\infty}{P} + \frac{\phi_\infty}{N_c}) + N_A\chi_{AP} + N\chi(1-\frac{N}{N_c}\phi_\infty)^2 , \tag{66}$$

$N_c = N_A + N$ is the diblock degree of polymerization.

The volume fraction ϕ_∞ of the copolymer A-N in the host matrix P controls the surface (interfacial) coverage σ (see Eq. 64) through its influence on μ_{bulk}. The segregation isotherm $\phi_\infty(\sigma)$, characterizing the relation between ϕ_∞ and σ, can be easily calculated numerically. In the limit of very small fraction ϕ_∞, segregation isotherm $\phi_\infty(\sigma)$ can also be expressed analytically for a wet (P<N) brush:

$$\phi_\infty(\sigma)\Big|_{wet} = (N\frac{\Omega}{a^3}\sigma)\exp\{5\frac{3^{1/3}}{2^{5/3}}N(\frac{1-2P\chi}{P})^{2/3}(\frac{\Omega}{a^3})^{2/3}\sigma^{2/3}$$

$$+\frac{N_A}{P}+\Delta_i-N_A\chi_{AP}\}$$

(67a)

and a dry (P>N) brush:

$$\phi_\infty(\sigma)\Big|_{dry} = (N\frac{\Omega}{a^3}\sigma)\exp\{\frac{9}{2}N(\frac{\Omega}{a^3})^2\sigma^2+\frac{N_c}{P}+\Delta_i-N\chi-N_A\chi_{AP}\}$$

(67b)

It is evident from these formulae that the areal density of segregated diblocks depends mainly on the *(mean field) adsorption parameter* $\beta'=N_A\chi_{AP}-\Delta_i$, describing the overall energy gain of a diblock at the interface as compared to the bulk. β' is equal to the height of the well of the potential U, affecting "anchor" blocks (see Fig. 33c). The linear part of Eq. (67) is yielded by the entropic term (the second in Eq. 57) due to the restriction of segregated copolymers to two-dimensional interface. The first term in the exponent of Eq. (67) originates from elastic energy due to brush stretching, supplemented eventually by osmotic contributions. The above expressions are obtained within the de Gennes-Leibler theory, which does not neglect the entropic term due to two-dimensional translational freedom of diblocks localized at the interface, as was postulated in original work by Leibler [231]. Therefore the regime of low surface coverage $\sigma(\phi_\infty{\to}0)$ is described properly here, contrary to what has been claimed [250, 253, 261] for the original Leibler theory.

A more detailed, although less clear, insight into the physics of segregated copolymers might be obtained in the frame of the self consistent mean field (SC-MF) numerical approach, developed by Shull and coworkers [236–238, 256, 260]. In this case an analogy between the path of a particle diffusing in a curvilinear fashion and a polymer coil is used. The chain is described by probability distribution functions [236] determining local polymer concentration. The interfacial free energy and the polymer profiles in the interfacial region are determined by solving a set of modified diffusion equations for such distribution functions in mean fields being themselves functions of the local composition [237]. While the SCMF numerical approach can be applied to each specific segregation case separately [242, 250, 252, 260–264], a universal solution has been obtained by Shull [237] for brushes composed of end-adsorbed N-mers in a chemically *identical* ($\chi{\approx}0$) high molecular weigh (P>N) P-mer melt [251, 253, 254, 256]. In this *asymptotic dry brush limit* the interfacial coverage by N-mer brushes is only a function of their radius of gyration $R_g(N)$ and an adjusted bulk chemical potential$<\mu_{bulk}>$:

$$\frac{<\mu_{bulk}>}{k_BT} = \frac{\mu_{bulk}}{k_BT} - \Delta_i - 1.1\ln(R_g(N)/w_A),$$ (68)

where the last component is the entropic penalty associated with the confinement of anchoring groups or segments to the interfacial region of width w_A (see Fig. 33) [237]. An inspection of Eq. (68) together with Eq. (66) yields that the only adjustable parameter of the chemical potential$<\mu_{bulk}>$would be the *SCMF adsorption parameter β*:

$$\beta = N_A\chi_{AP} - \Delta_i - 1.1\ln(R_g(N)/w_A) = \beta' - 1.1\ln(R_g(N)/w_A)$$ (69)

β turns out to be a central parameter in describing the thermodynamics of brush formation.

While in the de Gennes-Leibler model the brush "anchors" are strictly confined at the interface, the SCMF numerical approach by Shull allows for their distribution in a finite region adjacent to the interface. Similar assumption was made in the frame of the SCMF approach by Semenov [235], who derived the analytical expression for the segregation isotherm in the dry brush regime (P>N).

4.2
Review of Experimental Results

4.2.1
Conformation of a Polymer Brush

Numerous experimental studies have dealt with the problem of the conformation of N-mer brush layers immersed in P-mer melts $(P>N^{1/2})$, and created by diblock copolymers [240, 241, 261] or other end-functioned polymers [242–245, 249, 262, 264, 265]. The volume fraction vs depth profiles $\phi(z)$ of end-segregating polymers (as in Figs. 36), obtained in the course of these studies, yield the surface coverage σ. σ is related to the surface excess z^* as

$$\sigma = \frac{z^*a^2}{N_c\Omega} = \frac{a^2}{N_c\Omega} \int_0^{z(\phi_\infty)} [\phi(z) - \phi_\infty]dz$$ (70)

$z(\phi_\infty)$ is the depth at which the plateau in composition with bulk concentration ϕ_∞ is reached.

Spatial features of the brush profile are observed with the precision determined by the resolution ρ of depth profiling techniques used. The resolution ρ, described as a half width at half maximum (HWHM) of a Gaussian function, should be at least comparable with the unperturbed dimension of the brush, characterized by its radius of gyration $R_g(N)$. Therefore nuclear reaction analysis [19] (NRA, ρ=8 nm), secondary ion mass spectroscopy [26, 27] (SIMS, ρ=

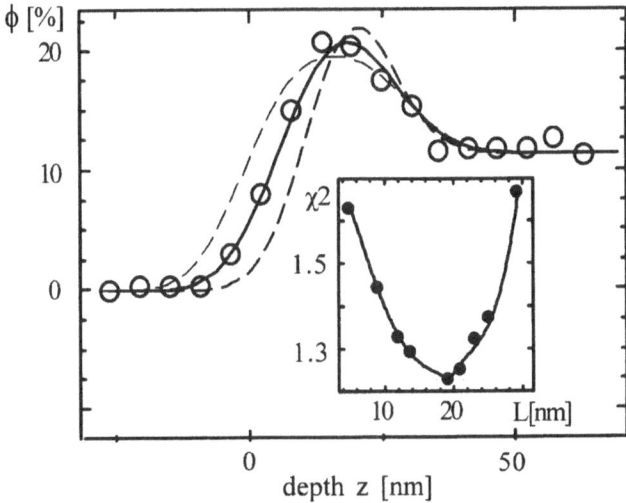

Fig. 36. Typical [241] volume fraction vs depth $\phi(z)$ profile of PI-dPS (N=893) copolymers in a PS (P=88) matrix, after 24 h of annealing at 180 °C. The NRA profile (O), corresponding to the brush layer created by diblocks at the free PS surface (z=0), is approximated by the convolution of the method resolution (ρ=8 nm) with the top hat-like function of Fig. 33b with different width L but constant z^*. The *inset* shows the variation of the fit parameter (mean square of residuals) with L. Best fit obtained for L=19(2) nm generates the profile marked by a *solid line* in the main figure. For comparison profiles calculated for L= 10 and 30 nm are shown by *broken lines*

5 nm), and neutron reflectivity [28] (NR, ρ=1 nm) can be, in principle, used to study (polystyrene) brushes with N larger than 855, 334, and 13, respectively.

The resolution $\rho \approx R_g(N)$ already allows us to obtain an explicit measure of the brush layer thickness L (see Fig. 36). In this case the simplest step-function profile $\phi(z)$ with constant composition in the brush layer region is assumed (see Fig. 33). While the de Gennes-Leibler model assumes all end-attached chains to stretch at the same distance z=L from the interface, the situation with a lower free energy is conceivable [226, 228] characterized by the non-uniform stretching and the total brush concentration decreasing with z. Measurements performed with higher resolution reveal [242, 243, 261, 264] the profiles $\phi(z)$ of the stretched brushes which might be approximated by an error function [266]:

$$\phi(z) = \frac{1}{2}[(\phi_{max} + \phi_{\infty}) - (\phi_{max} - \phi_{\infty})erf(\frac{z-H}{w\pi/\sqrt{6}})] \tag{71}$$

where the offset position H is used as a measure of the brush height H=L [226, 228, 242]. Such profiles, also yielded by SCMF calculations, are typical for brush chains anchored to neutral interfaces. Visible deviations from Eq. (71) are observed experimentally [262] when the interface attracts preferentially not only the anchoring moiety A of the N-mer brush but also the host melt P or the N-mer itself. Also the profile $\phi(z)$ of the unstretched brush [244] (with overlapped chains attached to neutral or attractive interface) is found to differ from the form proposed by Eq. (71) or by any function with a maximum concentration ϕ_{max} at (or very close to [261]) the interface. In this case also SCMF fails and the profile peaked at $z \approx R_g$ is interpreted [244, 265] by single chain ideal Gaussian statistics modified additionally for attractive interfaces.

First experiments which focused on the variation of the conformational properties have been performed by Brown et al. [240], who studied the role of the interactions between matrix and brush polymers (*enthalpy driven brush swelling*, see Eq. 59). They used a series of polystyrene (PS)-poly(methyl methacrylate) (PMMA) symmetric diblock copolymers with different blocks labeled by deuterium, placed at the interface between PMMA and poly(2,6-dimethylphenylene oxide) (PPO) homopolymers. A double brush layer was created with PMMA blocks dangling into (neutral) PMMA homopolymer and PS blocks immersed in favorably interacting PPO melt ($\chi = \chi_{PS/PPO} < 0$). The SIMS profiles obtained showed that the PS side of the block copolymer is stretched by at least a factor of 2 with respect to the PMMA side.

Recently Clarke et al. [243] reported on the extended series of experiments tracing the same effect for a brush exposed to a larger variety of different melt matrices. To obtain almost constant interface coverage, they have used a method allowing them for a permanent grafting of end-functioned dPS chains (N=712) into a silicon substrate. dPS brush chains were observed to be strongly swollen ($L \approx 3R_g$) when immersed in favorably interacting matrix polymers (such as poly(vinyl methyl ether) PVME or a poly(phenylene oxide)/PS mixture with $\chi < 0$) and to be collapsed when exposed to immiscible polymer melts (such as PMMA or polybutadiene (PBD) with $\chi > 0$). With increasing interaction parameter χ, the penetration of the host matrix P-mers into N-mer brush is limited and the width of the interface between brush layer and the bulk becomes narrower.

Practically all observations [241–243, 245, 249, 261, 262, 264] of the *entropy driven brush stretching* (see Eq. 59) have been performed for polystyrene brushes immersed in an identical, except for the isotopic status, PS matrix. The change in the conformational properties of the brush has been observed as a function of surface coverage σ [241–243, 245, 249, 261, 262, 264] and/or host matrix degree of polymerization P [241, 243]. These studies were initiated in 1992 by us [241] and two other groups [249, 262]. Here we resume the results of these reports focusing on the verification of the schematic conformation diagram (lnP/ln N vs lnσ/ln N) predicted by theory [232, 239, 267] and presented in Fig. 35.

Two relatively similar brushes – PI-dPS(N=893) at the free surface of the PS matrix (P=88–3173) and (COOH)dPS (N=929) grafted onto silicon substrate in the PS matrix (P=6442) – were studied by NRA [241] and SIMS [249], respectively. The NRA data were fitted with a simple top hat brush profile (Figs. 33 and 36) and its width was taken as a measure of the brush height L. In the other study, the mean brush height L was approximated by the half-width of the brush layer as measured by SIMS. The variation of L with the surface coverage σ is shown in

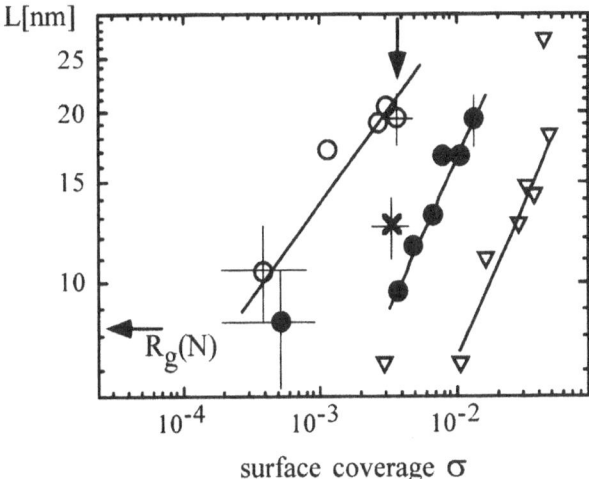

Fig. 37. Variation of the brush height L with the surface coverage σ, as evaluated by NRA (Fig. 36) [241] for PI-dPS (N=893) copolymers segregated at a free surface of a PS matrix with P=88 (○), 495 (✕), 3173 (●); and as determined by SIMS [249] for (COOH)dPS (N= 929) grafted onto the silicon substrate of PS(P=6442) matrix (∇). *Solid lines* mark the variation L∝$q^{1/3}$ concluded for the data (○) and L∝$q^{1/2}$ for the data (●) and (∇). A *horizontal arrow* shows the brush radius of gyration, while a *vertical arrow* marks the equi-σ situation described in details by Fig. 39 (here $\Omega=a^3$)

Fig. 37 on a double logarithmic plot for different homopolymer matrices (P=88–3173 [241] and 6442 [249]). At low coverage σ all data sets approach a value of L which is close to $R_g(N)$. With the surface coverage increasing, the brush height L increases in a manner which depends on the degree of polymerization P of the host matrix homopolymer.

First, at higher P (see Fig. 37) the onset of the brush swelling is shifted to higher σ values. This trend is due to the excluded volume interactions being progressively screened out. The onset of stretching is expected at a P-dependent crossover coverage $\sigma_2 \cong PN^{-3/2}(1-2P\chi)^{-1}$ for P<N and at a constant coverage $\sigma_5 \cong N^{-1/2}$ for P>N (see Fig. 35). Such a prediction is in accord with observations made exclusively for the NRA data. In spite of the fact that NRA and SIMS results were obtained for an N-mer brush attached to different interfaces and observed with a different resolution, the comparison of these results might suggest that the influence of the homopolymer degree of polymerization on the onset σ value might be still effective for P>N.

Second, the variation of the brush height L with the surface coverage σ is described by power law L∝σ^q with different q values depending on P. The data corresponding to the brush immersed in P=88 [241] matrix can be described by q= 1/3 while that corresponding to P=3173 [241] and 6442 [249] by q=1/2 (see Fig. 37). The range of coverages σ and matrices P used, corresponding to these two experiments, is marked on the ln(σ)/ln(N) vs ln(P)/ln(N) plot in Fig. 38. The

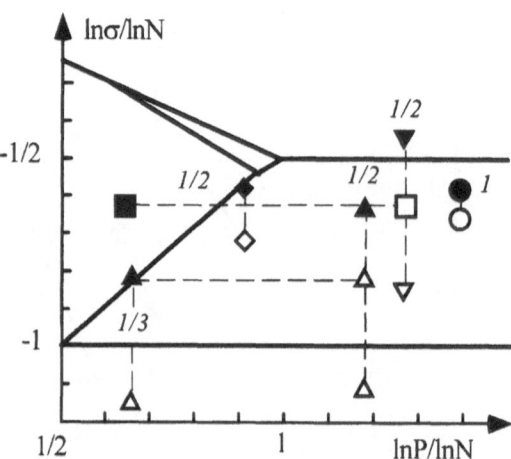

Fig.38. The comparison of the predicted by theory conformation diagram (see Fig. 35) with the experimental observations of brush stretching made for *equi-P situation* (($\triangle$, $\blacktriangle$) for [241], (∇, $\blacktriangledown$) for [249], ($\diamondsuit$, $\blacklozenge$) for [245], and ($\bigcirc$, $\bullet$) for [264]) and for *σ=const* (($\triangle$, $\blacktriangle$) for [241] and ($\square$, $\blacksquare$) for [243]). *Dashed lines* terminated by symbols mark the σ or P ranges. The progress of a brush swelling goes from *open to filled symbols*. Numbers are the observed exponents q of the $L \propto \sigma^q$ variation

NRA [241] and SIMS [249] data are marked by $\triangle$, $\blacktriangle$ and ∇, $\blacktriangledown$, respectively. The direction of observed increase in brush swelling goes from an open to a filled symbol. Observed q values are also marked in Fig. 38. The variation $L \propto \sigma^{1/3}$ indeed occurs very close (for $\ln(P)/\ln(N)=0.66$) to the wet brush "III wet" region, where it is expected. However, the power law $L \propto \sigma^{1/2}$ determined for $\ln(P)/\ln(N)=1.19$ and 1.28 is located away from the predicted region "III t", characteristic for a wet brush with a small excluded volume parameter $\Omega(1-2P\chi)/P$.

Before discussing this situation we mention three other experimental results. Reiter and coworkers [245] have recently observed, in NR experiments, the variation $L \propto \sigma^{1/2}$ of the brush height for PS (N=817) chains end-grafted onto a silicon substrate and immersed in dPS (P=464) matrix. The $\ln(P)/\ln(N)$ value=0.92 and the range of obtained surface densities σ, marked by symbols $\diamondsuit$, $\blacklozenge$ in Fig. 38, locates this observation very close to the predicted q=1/2 "III wet" region. On the other hand, the power law $L \propto \sigma^1$ (characteristic for the dry brush "III dry" regime) has been observed by Clarke et al. [264] for $\ln(P)/\ln(N)=1.40$. Their NR studies have been performed for (COOH)dPS (N=712) brush in the PS (P=4815) host matrix with the coverage range σ marked by $\bigcirc$, $\bullet$ in Fig. 38. These observations have been reconfirmed for the same system but with the lower value $\ln(P)/\ln(N)=1.29$ [242] (for the sake of clearness not marked in Fig. 38).

The surface densities σ predicted for the $L \propto \sigma^{1/2}$ region "III t", are almost reached in all three experiments [241, 245, 249], where this scaling law is observed. Values of numerical factors used in the theoretical approach (see Sect. 4.1) leading to the brush conformation behavior are assumed arbitrary

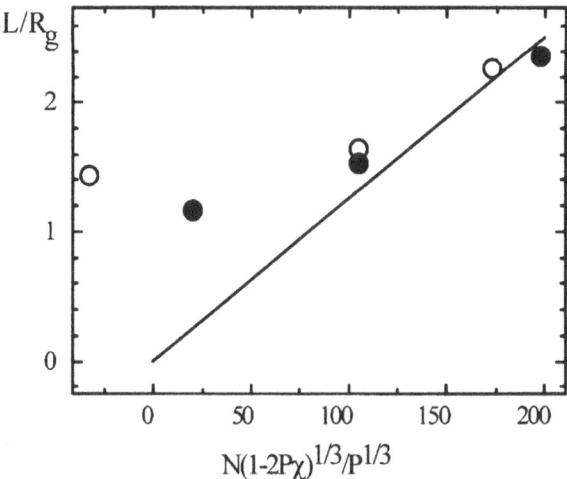

Fig. 39. The observed variation of the brush height L as a function of $N(1-2P\chi)^{1/3}P^{-1/3}$ for σ = const (Eq. 59). Points marked by ● and ○ correspond to the data of [241] (Fig. 37) and [243], respectively

(e.g., stretching is defined for $L>R_0(N)$ and not for $L>R_g(N)$). Therefore some shift in $\ln(P)/\ln(N)$ is expected for the exact location of the "III t" region in the conformation diagram plane $\ln(\sigma)/\ln(N)$ vs $\ln(P)/\ln(N)$ [232]. On the other hand, the assumptions leading to Eq. (61) are valid only if the chains of the host matrix P chains are present in the wet brush layer. This condition is equivalent to a relatively low volume fraction of brush N-mer chains in the brush layer. Φ is indeed less that 0.45 and 0.76 in experiments reported by Budkowski et al. [241] and Reiter et al. [245], respectively. However, Φ is very close to 1 for the situation described by Zhao et al. [249]. Therefore the apparent exponent q=1/2 in the latter case may merely indicate a transition region between the unswollen q= 0 (at low σ values) and dry brush q=1 (at higher σ values) regimes expected for high molecular weight matrix (P>N). Indeed the "pure" q=1 behavior is observed for a slightly higher $\ln(P)/\ln(N)$ ratio [242].

Now we compare the variation of the brush thickness L with P, as predicted by Eq. (59), with the available data. The first experimental study of this problem [265], performed by SIMS for a series of very short (as compared to a standard depth resolution of this technique) (COOH)dPS (N=125–413) end-anchored chains annealed at different temperatures and with different matrices used (T= 108 °C for P=63, and 160 °C for P=4460), revealed identical unswollen brush conformation for both used matrix molecular weights. The variation of the brush height L with P, at a *constant coverage* σ, has been observed by two other groups in experiments performed later [241, 243]. On the basis of the cumulative data for the PI-dPS (N=893) brush in the PS host matrix (P=88, 495, and 3173), presented in Fig. 37, we were able to achieve an equi-σ situation ($\sigma=3.7 \cdot 10^{-3}$, see arrow in Fig. 37 and corresponding symbols ($\triangle$, $\blacktriangle$) in Fig. 38 as described in

[241]. Recently, Clarke et al. [243] also reported on the σ=const regime, obtained for end-functioned dPS chains (N=712) permanently grafted to a silicon substrate and exposed to PS matrix with P=67–4815 (marked by symbols $\square$, $\blacksquare$ in Fig. 38. The comparison of the theoretical prediction $L \propto N(1-2P\chi)^{1/3}/P^{1/3}$ with the results of these two reports (marked by $\bullet$, $\bigcirc$ for [241] and [243], respectively) is presented in Fig. 39. We can see at once that while the data points with moderate P values seem to follow the theoretical prediction (marked by the solid line), the data pair corresponding to the highest P values is clearly different, apparently out of the wet brush ("III wet") regime. We notice also, that for high P-mers the factor $(-2P\chi)$, although originating from small isotopic interaction parameter χ, is large enough to make the virial coefficient $(1-2P\chi)\Omega/P$ equal to zero and to allow the "III t" regime ($L \propto \sigma^{1/2}$) to occur (see Fig. 35). In any case Fig. 39 clearly shows the swelling of the brush by the low molecular weight matrix. In addition to this change in the brush height L, Clarke et al. [243, 264] also reported the increase in the interfacial width (expressed by w in Eq. 71) between the brush and the matrix which occurs with decreasing P.

4.2.2
Diblock Segregation Modified by a Homopolymer Chain Length

Experimental observations of diblock segregation have been reported since 1990 [237, 240, 246, 260]. Very soon more extended studies on molecular weight dependence of segregation have appeared, initiated by Green and Russell [248], Dai et al. [250] and by us [251]. The first and last groups studied diblock copolymer A-N attached by their "anchor" block A to an interface created by P-mer host matrix with vacuum and/or substrate. The second group studied diblock copolymer A-N positioned at the interface between homopolymers A and N. Before referring to this double-brush situation we describe here observations done for a single brush layer.

Green and Russell [248] used a deuterated, symmetric (N_c=262) PS-PMMA copolymer immersed in PS homopolymer matrix with a varied degree of polymerization P. They observed identical segregation isotherms for the studied range P=1000–17,300. Their conclusion is in accord with other reports [250, 253] confirming that the segregation isotherm is independent of matrix molecular weight for $P>3N$. This condition defines practically the *dry brush regime* expected on the basis of the theory. Indeed, the segregation isotherm formula (Eq. 67b) does not depend on P when P is large (and N_c/P small).

Another range of matrix molecule sizes P=88–3173 was used in our study [251] on PI (polyisoprene, N_A=114)-dPS (N=893) diblock copolymer segregating to interfaces created by polystyrene P-mer with vacuum and silicon substrate. The used PS molecular weights covers the wet and dry brush regime. In Fig. 40a we present typical composition-depth profiles of PI-dPS obtained at the vacuum ("external") interface of PS host matrix with P varied (P=88, 495, and 3173), but constant bulk diblock concentration ϕ_∞=3.2(5)%. The surface peak and a related surface excess z^* (and coverage σ) increases with P. This is even

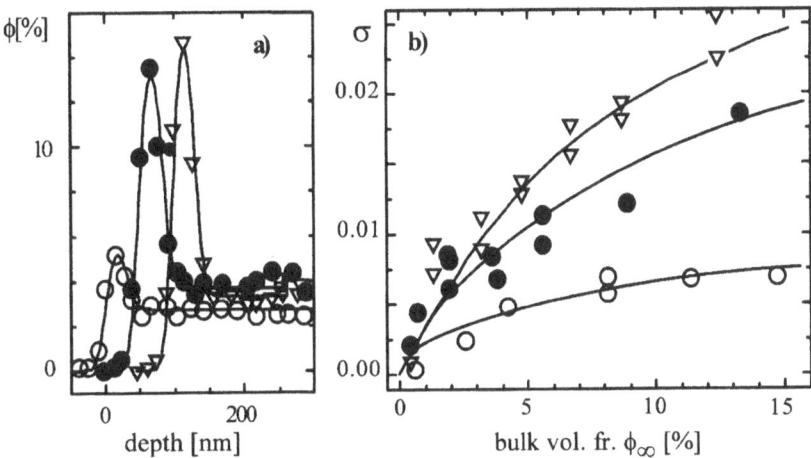

Fig. 40.a Composition-depth profiles of PI-dPS (N=893) copolymer at the vacuum interface of different PS homopolymers (P). **b** The corresponding segregation isotherms [251]. Situations corresponding to different matrices are marked by: ○ for P=88, ● for P=495 and ▽ for P=3173 (results are insensitive to temperature modification [251])

more evident in Fig. 40b, where segregation isotherms $\phi_\infty(\sigma)$ are plotted for P= 88 (○), 495 (●), and 3173 (▽). This behavior can be easily explained by Eq. (67a) expressing $\phi_\infty(\sigma)$. As a result of the trade between elastic energy and excluded volume interactions (modified by P), these two contributions result in the "bare" brush energy, dependent on $(\sigma/P)^{2/3}$ and determining the segregation properties (Eq. 67a). An increase in coverage σ is expected for increasing P for both bulk concentration ϕ_∞ and adsorption parameter β' ($=N_A\chi_{AP}-\Delta_i$) kept constant. Surface areal density σ is independent of P in the dry brush limit.

The segregation isotherm formulas (Eq. 67) can be used to fit the data of Fig. 40b. We use the wet brush expression for P=88 and 495 and the dry brush expression for P=3173. They reveal that the real situation is more complex than sketched schematically above. Namely, the adsorption parameter β' turns out [251] to be P-dependent, as we have $\beta'(88)=7.2$, $\beta'(495)=5.9$, and $\beta'(3173)=5.2$. This changes the difference between the segregation for P=88 and 3173 as compared with the prediction made for constant β'. Let us assume that the energy Δ_i, associated with the adsorption of the PI moiety of the diblock to the vacuum interface, remains unchanged while P varies. In this case the observed $\beta'(P)$ relation might suggest that the magnitude of the segment PS – segment PI interaction parameter χ_{AP} is a decreasing function of P. This is consistent with the results of the small angle X-ray scattering study of PI-PS diblocks in PS matrices [268], where a marked increase in χ_{AP} was observed with decreasing P. Detailed comparison cannot be made because different P and ϕ_∞ values were used in the two studies, although similar values and variations of χ_{AP} were determined.

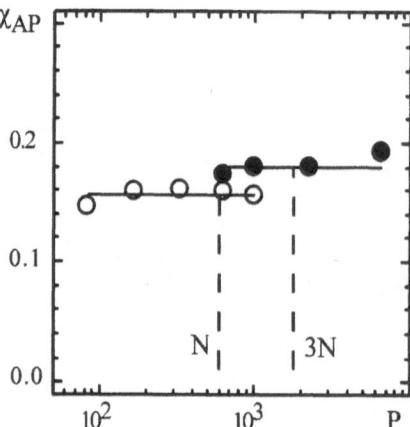

Fig. 41. The segmental interaction parameter χ_{AP} between the anchoring block (PVP) of the dPS (N=597)-PVP copolymer and the seven different P-mer PS host matrix molecules, as obtained from fitting the segregation data by Dai and Kramer [253] with mean field models describing dry-wet (○) and dry-dry (●) double brush layer at the PS//PVP interface

The P-mer size dependence of the parameter χ_{AP} fitted to the segregation data of PI-dPS diblocks [251], was related to the physically occurring behavior. However, also another opinion has emerged [250, 253, 261, 263]. It expects *strong* P-variation of χ_{AP} determined based on segregation isotherms, as originating from the shortcomings inherent in the de Gennes-Leibler model. This opinion is based, however, on a relatively limited analysis [250] performed for only two, very different, P-mers (P<N and P>N). In addition, two different (simplified and complete) forms of de Gennes-Leibler approach were used [250] for long (P>N) and short (P<N) P-mer.

To verify this opinion we use the extended data set [253] (seven different P-mers) available for the *double brush* layer created by dPS (N=597) – PVP (poly(2-vinylpyridine), N_{PVP}=59) diblock at PVP (P_{PVP}=2000)/PS (P=86–6442) interface. While the PVP blocks dangling into PVP homopolymer create a dry brush layer, the dPS blocks immersed in PS matrix form wet- or dry-brush. The free energy G of the dry-wet or dry-dry double brush layer may be written as a superposition of Eqs. (63) and (60) or (63), respectively. Then the expressions for the segregation isotherm $\phi_\infty(\sigma)$, similar to Eq. (67), are obtained. Finally, the segregation isotherms [253] (reported for various P) are fitted to yield the parameter χ_{AP} corresponding to dry-wet (○) or dry-dry (●) double brush model (see Fig. 41). We can see that only a relatively small change in χ_{AP} may be advocated within each model, while a jump in the mean value $\langle\chi_{AP}\rangle$ is observed: from 0.155(5) for a dry-wet brush to 0.180(6) for dry-dry double brush layer. Both values differ significantly from the value 0.109(4) reported by SCMF calculations [253]. Smaller deviations of the value χ_{AP} yielded by SCMF and de Gennes-Leibler models are expected for a *single* brush layer, as would be concluded based on the comparison done for the asymptotic dry brush limit and presented in the next paragraph.

4.2.3
Molecular Architecture of a Diblock and Its Segregation Properties

Experimental research, focused on the influence of the diblock architecture on its segregation properties, has been initiated by Shull et al. [260] and continued later for identical and different system by Dai and Kramer [253] and by us [254]. Diblock architecture can be characterized by its asymmetry $r_c=N_A/N_c$ defined as the ratio of degree of polymerization describing the "anchor" block (N_A) and the whole copolymer ($N_c=N_A+N$).

Shull et al. [260] studied dPS-PVP copolymers with $r_c=0.12$ ($N_c=1473$) – 0.26 ($N_c=480$) immersed in PS host matrix($P=6000$) and segregating to its interfaces with vacuum and with PVP homopolymer. For low bulk concentrations ϕ_∞, the copolymer excess has been observed only at the PS//PVP interface. For ϕ_∞ exceeding a critical value ϕ_{cmc}, the segregation also started at a free surface, while that to PS/PVP interface was increased considerably outside a short transition region. These observations were explained by the formation of copolymer aggregates (*micelles*) as confirmed later by electron microscopy [269]. A micelle consists of an inner region built of "anchor" blocks A (here PVP), and an external region created by N-mer block brushes (here dPS) exposed to homopolymer matrix. Due to the higher surface tension of PVP as compared to dPS, no segregation of individual dPS-PVP diblocks to the free surface is expected. However the micelles (present for $\phi_\infty>\phi_{cmc}$), where PVP core is encapsulated in dPS corona, behave effectively as large dPS molecules attracted to the free surface. The micelle formation is associated with halted segregation of individual copolymers, followed by the onset of interfacial segregation of micelles themselves.

Critical micelle concentration ϕ_{cmc} is expected to decrease strongly with diminished diblock asymmetry r_c as low r_c values favor easier creation of highly curved micelle interfaces. Theory of micelle formation [231, 260] also indicates that the overall copolymer degree of polymerization N_c, as well as the "anchor"-homopolymer interaction parameter χ_{AP}, have to be considered to explain properly the onset of micelle segregation as observed by Shull et al. [260]. Using this theory, experimenters are able to choose systems where only individual copolymers segregate.

Dai and Kramer [253] studied free dPS-PVP diblocks, with $r_c=0.09$ ($N_c=656$) – 0.27 ($N_c=164$), immersed in a high molecular weight PS host matrix and creating a double dry brush layer at the PS/PVP interface. The segregation isotherm data were analyzed on the basis of the predictions yielded by self consistent mean field (SCMF) theory for *asymptotic dry brush limit* (see Eqs. 68 and 69). In this approach segregation is characterized by the interfacial excess $z^*_N=(1-r_c)z^*$ of the N-mer brush forming blocks, rather than by the excess z^* of the whole copolymer The normalized excess $z^*_N/R_g(N)$ depends only on the effective bulk chemical potential $<\mu_{bulk}>$ (Eq. 68) adjusted using SCMF adsorption parameter

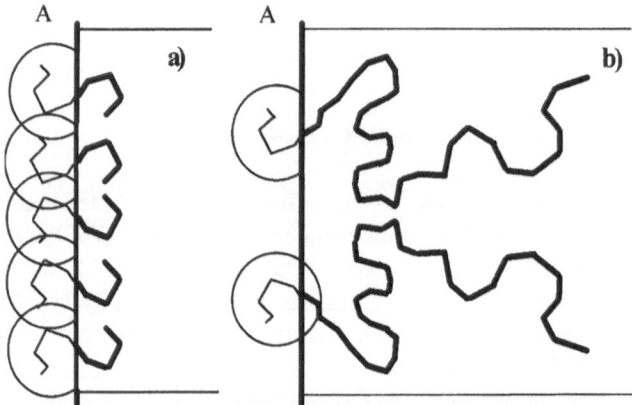

Fig. 42a,b. Cartoon to illustrate the origin of different values of the adsorption parameter β' ($=N_A\chi_{AP}-\Delta_i$) for the symmetric and asymmetric diblocks; **a** the short symmetric copolymers segregate to the (vacuum) interface of the homopolymer matrix at areal density σ, high enough to cause the overlap of anchoring (A) blocks. The interfacial energy reduction Δ_i, due to the positioning of an additional "anchor" block at the interface, is then lower than for the case where anchoring (A) blocks do not overlap. **b** The latter applies for the long asymmetric copolymers

β (Eq. 69). The relation $z^*_N/R_g(N)$ vs $<\mu_{bulk}>$ is described by a master curve calculated numerically by Shull [237]. For a *double brush* layer created by dPS-PVP copolymer at PS/PVP interface, a few modifications in this approach were suggested [253]. Firstly, it was noticed that dPS-PVP diblocks present at the PS/PVP interface do not change the extent of PS/PVP interactions at the interface, hence the Δ_i term in β is equal to *zero* (see Eq. 69). Secondly, the joints between copolymer blocks were assumed to be confined at the PS/PVP interface characterized by its width ($w_A=w$). Finally, the relevant spatial scale characterizing individual unstretched double-brush would be the copolymer radius of gyration $R_g(N_c= N+N_A)$ rather than $R_g(N)$. With these changes, the relation $z^*/R_g(N_c)$ vs $<\mu_{bulk}>$, yielded by experimental data, followed the SCMF master curve well [253]. Thus we can predict the segregation of any copolymer A-N to an interface A/P between homopolymers with high molecular weights if only the molecular architecture (r_c, N_c) of the diblock is known.

A more complex situation is present for free diblock copolymers A-N segregating from a high molecular weigh homopolymer P to its interfaces with a vacuum (free surface) or a substrate. Here a *single brush* layer is created by N-mer blocks attached to the interface by "anchor" blocks A. They are driven by lower interfacial energies of A as compared to P segments and by the enthalpic penalty of A blocks embedded in P-mer matrix (Fig. 33). For low areal density σ the anchoring blocks create separate patches which transform, for higher surface densities σ, into a continuous pancake film (see Fig. 42). The interfacial energy reduction Δ_i, due to the positioning of an additional "anchor" block A at the interface, is *higher* for the P-mer interface not completely covered by A segments

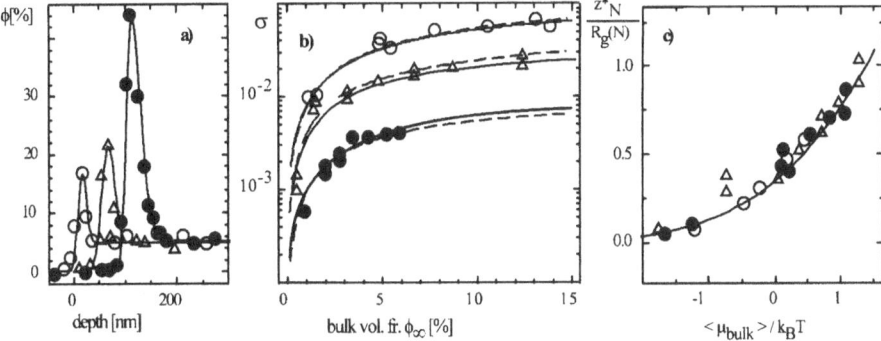

Fig. 43.a Composition-depth profiles [251, 254] of PI(N_A=114)-dPS diblock copolymers with various N-mer brush forming dPS blocks (N=89 (○), 893 (△), and 9551 (●)) at the free surface of the high molecular weight P-mer (P>2.9 N) PS host matrix. **b** The corresponding segregation isotherms analyzed with mean field (*solid lines*, Eq. (67b)) and self consistent mean field (SCMF, *dashed lines*) approaches. **c** Normalized interfacial segregation of dPS blocks $z^*_N/R_g(N)$ as a function of the normalized chemical potential of Eq. 68 [254]. The *solid line* shows the SCMF master curve [237]

(Fig. 42b). This is because the effective exposed area of the P-mer matrix interface, as well as the related interfacial energy, is considerably reduced by each segregated diblock. For high interfacial coverage σ only a small exposed host matrix area is left to cover and Δ_i is *smaller*. Thus in order to predict the diblock segregation properties, the role of diblock architecture on the Δ_i term in the adsorption parameter (β or β') should be verified.

This problem has motivated our studies [254] of a series of free PI-dPS diblock copolymers, with *identical* PI "anchor" block (N_A=114) and asymmetry ratio varied by two orders of magnitude, namely from r_c=0.01 (N_c=9665) to r_c=0.56 (N_c=203). This situation is exceptional as micelles are formed for similarly small r_c values for other diblocks (due to higher χ_{AP} [260]). PI-dPS diblocks segregate from a high molecular weight PS matrix to its interfaces with vacuum and Si substrate. These studies are illustrated here with composition-depth profiles of short (N_c=203, ○), intermediate (N_c=1007, △) and long (N_c=9665, ●) copolymers at the vacuum interface of PS homopolymer (with P>2.9 N) with identical bulk diblock concentration ϕ_∞=5.1(3)% (Fig. 43a). Corresponding segregation data are presented in terms of the density σ of diblock chains per unit area of the interface and plotted as a function of ϕ_∞ in Fig. 43b. We note the main qualitative features: the shortest symmetric diblocks (N_c=203, r_c=0.56, ○) segregate to an interfacial density that is much higher (by over an order of magnitude) than that of the longest asymmetric (N_c=9665, r_c=0.01, ●) ones for a given volume fraction ϕ_∞ of the copolymer. After an initial rapid rise, the interfacial density of the brushes appears to increase slowly with ϕ_∞. These segregation isotherms σ(ϕ_∞), obtained experimentally, are well described (see solid lines in Fig. 43b) by the mean field formula (Eq. 67b) with adsorption parameter β'=$N_A\chi_{AP}-\Delta_i$=5.2 for the longest and intermediate diblocks as well as 3.6 for the shortest copolymer.

We attribute the lower β' value obtained for shortest copolymer to the much reduced amplitude of the Δ_i term as compared to that of other copolymers. This is due to different morphology of PI anchoring diblocks at the surface as suggested above (Fig. 42). PI diblocks start to overlap for the critical areal density $\sigma_c = a^2/(\pi R_g^2(PI)) = 2.10^{-2}$. For densities lower than σ_c, i.e., for long and intermediate copolymers, isolated PI patches are expected at PS surface and hence larger amplitude of the Δ_i term is anticipated. On the contrary, continuous PI pancake and reduced Δ_i is predicted for shorter diblocks segregating with $\sigma > \sigma_c$.

The segregation data were also analyzed with the results of the SCMF approach for the asymptotic dry brush limit. Plotting the normalized surface excess $z^*_N/R_g(N)$ vs the effective bulk chemical potential $<\mu_{bulk}>$, we find that all segregation isotherms follow the SCMF master curve. This is shown in Fig. 43c, where the superimposed data for the short (O), intermediate (Δ) and long ($\bullet$) diblocks fit the results of SCMF calculations [237] (solid line) for the SCMF adsorption parameter β equal to 2.2, 2.7, and 1.5, respectively. The corresponding segregation isotherms predicted by SCMF are plotted in Fig. 43b (dashed lines). To enable the comparison of adsorption parameters yielded by mean field and by SCMF, we transform the SCMF parameters β into the corresponding mean field β'_{SCMF} ones. This is performed using Eq. (69), where the entropic penalty term was evaluated for the "anchor" blocks confined to the interfacial region with the width of $w_A = 0.9$ nm [251]. The obtained β'_{SCMF} values are in good agreement with the (mean field) β' parameters for intermediate ($\beta'_{SCMF} = 5.2$ vs $\beta' = 5.2$), long ($\beta'_{SCMF} = 5.1$ vs $\beta' = 5.2$), and short copolymers ($\beta'_{SCMF} = 3.4$ vs $\beta' = 3.6$). Slightly worse accord is obtained when the segregation isotherm predicted by mean field theory is not represented by Eq. (67b) but calculated numerically. This comparison suggests that both the mean field approach and the SCMF picture of single brush layers both yield equivalent values of the adsorption parameter $\beta' = N_A \chi_{AP} - \Delta_i$. While we cannot directly compare the χ_{AP} parameter yielded by the above approaches we may argue that χ_{AP} has similar values. This stems from the observation that the value of w_A (assumed to compare SCMF- and mean field results) is physically reasonable for a PI block confined to the free surface.

4.2.4
Adsorption from a Binary Mixture of Short and Long Diblock Copolymers

An understanding of the influence of molecular weight distribution on brush properties is very useful as polydispersity is often an unavoidable feature of polymer systems. In addition, a controlled molecular weight distribution of brush forming chains offers a possibility for tailoring the brush structure [270]. For these reasons, an increasing number of theoretical [270–273] and experimental [227, 255, 274–276] studies have dealt with this problem. Most of experiments considered a brush being formed from the binary mixture of short and long end-anchoring polymers immersed in a solvent. The first study of a mixed brush created at the interface of a polymeric matrix was reported by us [255]. Here we outline the main features of the results.

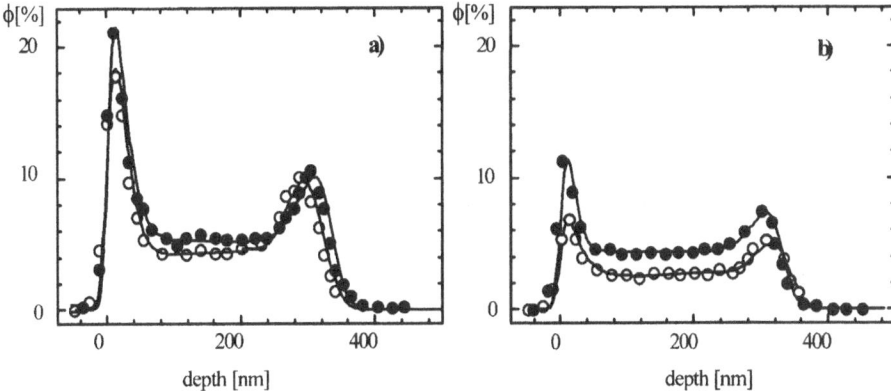

Fig. 44. Typical [255] composition-depth profiles of dPS blocks (after 3 days of annealing at 190 °C) in thin PS (m.w.=2.89×10⁶) homopolymer films with binary mixtures of short and long copolymers. Deuterated short copolymers dS=PI(N_A=114)-dPS(N=89) or their protonated analogs hS=PI (N_A=143)-PS(N=124) as well as long deuterated dL=PI (N_A=114)-dPS(N=9551) copolymers are used in pairs of samples with hS/dL (○ profiles) and dS/dL (● profiles) mixtures. The overall compositions of short (dS, hS) and long (dL) diblocks are: 2.1 and 6.7% for (a) as well as 4.2 and 3.3% for (b), respectively

We have studied [255] the segregation to the interfaces of PS (P=27,788) homopolymer from a binary mixture of short (designated "dS" with N_c=203, r_c= 0.56) and long (denoted "dL" with N_c=9665, r_c=0.01) PI-dPS diblock copolymers added to the PS host matrix. The adsorption properties of the pure components [254] of this mixture are described in the previous paragraph. To separate the surface and bulk contributions of the long and short diblocks we have used the hydrogenated analog (hS) PI-PS (N_c=267, r_c=0.54) of the short deuterated (dS) copolymer PI-dPS (N_c=203, r_c=0.56), both with almost identical segregation properties [116, 255]. We have used an *isotope contrast approach*, in which two series of samples were analyzed by NRA, tracing the composition-depth profiles of deuterated molecules only. The first series (filled circles in Fig. 44) consisted of PS matrices incorporating binary mixtures of deuterated long (dL) and deuterated short (dS) diblocks. The second series was composed of PS host matrix with binary mixtures of deuterated long (dL) but protonated short (hS) copolymers (open circles in Fig. 44). For each pair of samples: (dS/dL) and (hS/dL) the overall compositions were identical.

The isotope contrast method enabled us to extract directly, for each pair of samples, the interfacial excess of both short (z^{*S}_N) and long (z^{*L}_N) copolymers as well as their bulk concentrations ϕ^S and ϕ^L. We characterize the segregation by the interfacial excess $z^*_N=(1-r_c)z^*$ of the N-mer brush forming blocks rather than by the excess z^* of the whole copolymer. The results can be presented as a three-dimensional plot, where the dependence of the interfacial excess on the concentrations of the short (ϕ^S) and long (ϕ^L) copolymers is shown simultaneously. This is done in Fig. 45, characterizing the free surface of the PS matrix.

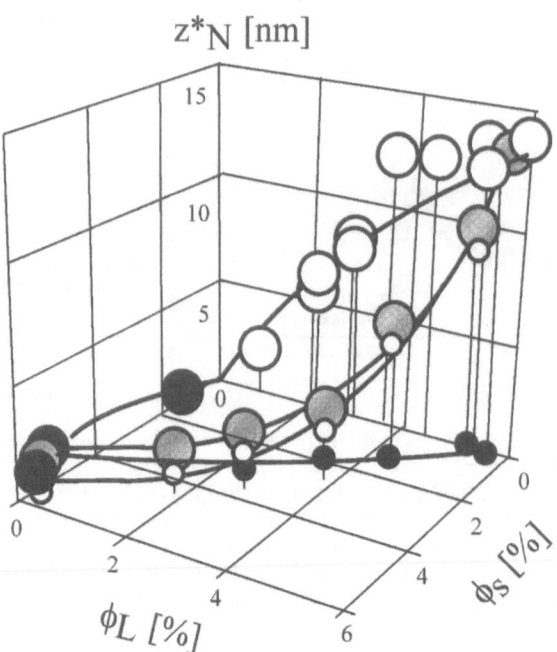

Fig. 45. Three-dimensional representation of the surface excess z^*_N of brush forming blocks at the vacuum/PS (m.w.=2.89×10⁶) interface from the binary mixture S/L of short (dS, hS) and long (dL) copolymers incorporated in the PS host matrix and described by the profiles as in Fig. 44. z^*_N is plotted as a function of bulk concentrations of short (ϕ^S) and long (ϕ^L) copolymers. While the cumulative excess z^*_N is marked by *shaded balls*, the contributions to z^*_N yielded by short and long diblocks are marked by *small black and white balls*, respectively. Data from single-component mixtures (see Fig. 43) are shown as *large black and large white balls* in the ϕ^L=0 and ϕ^S=0 planes

The overall interfacial excess of diblocks (shaded balls) is the sum of z^{*S}_N and z^{*L}_N (black and white small balls, respectively). The segregation isotherms of the pure short (black large balls) and pure long (white large balls) copolymers are plotted for comparison at the ϕ^L=0 and ϕ^S=0 planes, respectively.

The qualitative features of our results are already visible on composition-depth profiles shown in Fig. 44a,b. The behavior of short diblocks is characterized by the difference between (dS+dL) and dL profiles (marked by ● and ○, respectively) while that of long copolymers by dL profiles (○ points). The change in the surface excess of short diblocks, noticeable while comparing Fig. 44a and b, can be explained by the variation of the bulk concentration of short copolymers. On the contrary, a dramatic decrease in the excess of long copolymers cannot be explained by the change of bulk concentration of long diblocks alone. These features are more distinctly visible in Fig. 45. The comparison between segregation isotherm of the pure short copolymer case (large black balls) and the isotherm of short component of binary copolymer mixture (small black

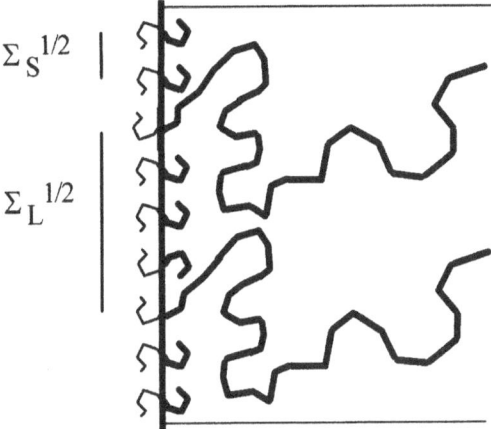

Fig. 46. Schematic illustration of the interfacial brush formed by the binary mixture of short symmetric (S) and long asymmetric (L) copolymers with identical anchoring blocks. Depicted situation corresponds to Figs. 44 and 45. Short and long components of the mixed brush are characterized by mean inter-anchor spacings $\Sigma_S^{1/2}$ and $\Sigma_L^{1/2}$ or equivalently by respective interfacial coverages σ^S and σ^L

balls) leads to the conclusion that the surface excess of short copolymers is hardly perturbed by the presence of longer diblocks. Similar comparison made for the pure long diblock case (large white balls) and the long component of binary diblock mixture (small white balls) reveals that the surface excess of the long copolymers is strongly reduced when short diblocks are added to the PS host matrix.

The above results can be explained with a simple approach, which is an extension of the model used to describe a monodisperse brush (Sect. 4.1). The simplicity of this approach is due to the specific structure of the studied brush formed by two different N-mer blocks with over a 100-fold size disparity (see Fig. 46). The inner brush layer, adjacent to the interface, and containing segments from both short and long chains, is very thin as compared with the outer layer containing segments of the long chains alone. Therefore we expect that there will be no interactions between short and long blocks over most of the trajectory of the latter. At the same time, the interfacial density of the short blocks in the inner layer is at least one order of magnitude greater than that of the long blocks. This implies that most of the free energy of the inner layer of the mixed brush is due to short blocks interacting with other short blocks. Thus the overall free energy F^{tot} of the mixed brush layer (segregated to the free surface of area A) is then approximated by an appropriate weighted sum of the free energies of the single-component brushes:

$$F^{tot} = \frac{A}{a^2}(\sigma^S G^S(\sigma^S) + \sigma^L G^L(\sigma^L)), \tag{72}$$

where superscripts refer to long (L) and short (S) copolymers. The form of G, the free energy per brush chain, is given by Eq. (63). It contains the interfacial energy reduction due to segregation of individual short (Δ_i^S) or long diblock (Δ_i^L). Δ_i^S differs from Δ_i^L as described in a previous paragraph. Interfacial densities of short (σ^S) and long (σ^L) chains are calculated using Eq. (70) from the surface excess z^* data.

The basic assumption of this approach is that the long and short copolymers segregated at the interface are in equilibrium with the long and short diblocks, respectively, incorporated (at concentration ϕ^S and ϕ^L) in the bulk of PS matrix film. This is expressed (see Eq. 64) by equating the chemical potentials of the copolymers at the interface (μ^S_{brush} and μ^L_{brush}) with those in the bulk (μ^S_{bulk} and μ^L_{bulk}):

$$\mu^S_{brush}(\sigma^S) = \mu^S_{bulk}(\phi^S, \phi^L) \qquad \mu^L_{brush}(\sigma^L) = \mu^L_{bulk}(\phi^S, \phi^L). \qquad (73)$$

The brush chemical potential of each individual component μ^S_{brush} or μ^L_{brush} depends only on the surface coverage of the same constituent, σ^S or σ^L, respectively. This is due to the simple form of the overall free energy of the mixed brush layer F^{tot} (Eq. 72). Individual brush chemical potential is easily evaluated using Eq. (65) for the individual free energy per brush chain specified by Eq. (63).

While the full relation for the chemical potential of each component in the bulk is given in [255] we discuss here only its essential properties. The above chemical potential is fully determined by bulk volume fraction of both components ϕ^S and ϕ^L, provided that relevant segment-segment interaction parameters (see Eq. 66 for a single component case) are known.

Using Eq. (73), the segregation isotherm of short $\sigma^S(\phi^S, \phi^L)$ and long copolymers $\sigma^L(\phi^S, \phi^L)$ can easily be predicted for a mixed brush case, when relevant adsorption parameters for single component segregation have been determined earlier. This is shown in Fig. 45 where the solid lines, calculated for the mixed brush layer with no adjustable parameters, provide a good quantitative fit to our data. Specific terms of the bulk chemical potentials [255] describing short and long copolymers ($\phi^L N^S / N^L$ for μ^S_{bulk} and $\phi^S N^L / N^S$ for μ^L_{bulk}) explain why the partial surface excess of each diblock is hardly or strongly perturbed by the presence of the another copolymer This is especially true for the large disparity N^L / N^S (>100) in the size of blocks forming the brush and belonging to long and short copolymers.

4.3
Summary and Conclusions

Diblock copolymers admixed to a homopolymer matrix segregate to its interface in a specific way. As well as other end-functioned molecules, diblock copolymers, attached by an anchoring block (or by a copolymer joint) to the interface, offer a convenient means for the realization of so-called polymer brushes. The structure of a polymer brush (characterized by its height and average composition) was predicted [230, 232] to be controlled by its interfacial coverage σ, the degree of polymerization of the brush (N) and matrix (P) molecules as well as by their segment-segment interaction parameter χ. The initiated [240, 241, 249] experimental studies on the brush conformation have confirmed this viewpoint (see Sect. 4.2.1). In particular they have shown that the swelling of the brush beyond its unperturbed size can by obtained increasing coverage σ [241, 242, 245, 249, 264] or magnitude of $(-\chi)$ [240, 243] or decreasing host matrix molecular size P [241, 243].

The interfacial segregation of diblocks is driven by the unfavorable interaction χ_{AP} between the anchoring block A of the copolymer and the matrix homopolymer P. This interaction is minimized by confining diblocks to the interface at the expense of free energy necessary to form the brush. Conformational studies (see Sect. 4.2.1) allowed us to verify two main terms of the brush chain free energy: a brush swelling is opposed by brush chain elastic energy but it is favored by free energy due to the mixing between the host matrix P-mer and brush molecules. As the last term scales as 1/P, the segregation ability of the diblock would be P-dependent (wet brush regime), except for very large P (say P>3 N, dry brush regime). This has indeed been observed [250, 251, 253] as described in Sect. 4.2.2. Segregation isotherms observed for varied P are all well depicted by the mean field approach and even better by the self consistent mean field (SCMF) picture (Fig. 41).

An additional factor driving the segregation exists whenever the anchoring moieties change interactions at the interface. The related reduction Δ_i in brush chain free energy may depend on the molecular architecture of diblocks (i.e., the size of blocks and of copolymer) as was recognized by our recent studies [254] described in Sect. 4.2.3. The results of these studies are analyzed with mean field and self consistent mean field approaches, both yielding identical adsorption parameters.

The influence of the molecular weight distribution of diblock copolymers on their segregating properties is considered in Sect. 4.2.4. It describes first experimental study [255] on a bimodal mixture of short symmetric and long asymmetric copolymers added to a polymeric matrix. Shorter copolymers were found to adsorb preferentially at the homopolymer interfaces in accord with brush formation observed from a solvent host matrix [274–276]. The mean field model is able [255] to predict the segregation isotherm of the bimodal mixture of copolymers, basing on single component adsorption data [254].

The determination of the conformational and segregation properties of polymer brushes, created by diblock copolymers, has triggered their application to more complex problems. Diblock copolymers have been used to increase adhesion [277] or to eliminate the interfacial tension [256] between immiscible polymers. They may also modify the surface induced mode [116] and the bulk mode [278] of the spinodal decomposition observed in homopolymer blends.

5
General Conclusions

In the present review we discuss experimental results concerning the interfaces observed in thin films composed of polymer mixtures, and compare them with theoretical expectations, based mainly on mean field models. The interfacial phenomena presented here are categorized according to their complexity. The behavior of the interface between two coexisting phases determines (Sect. 2) blends miscibility conditions. These, augmented by the knowledge of specific fields and effects relevant for an "external" interface of the binary blend, explain (Sect. 3.1) interfacial segregation. Similar picture, but involving both interfaces bounding a thin film, characterizes finite size effects (Sect. 3.2). Finally, we consider the interfacial brush layer typical for a ternary system (Sect. 4) with two types of molecules: homopolymers and diblock copolymers formed by two immiscible chains joined by a covalent bond.

The conclusions concerning each specific category of the studied phenomena follow each individual section. Here we mention briefly only some general points.

A mean field approach offers a good, clear frame to depict, in a consistent way, all various interfacial phenomena considered here. It covers simultaneously the *basic features* of their physics.

In addition, theoretical models extended beyond the standard mean field or self consistent mean field (for systems with sharp concentration gradients) approach can yield predictions which are in *quantitative* accord with experimental results. For instance, the inclusion of the concept of capillary wave excitations [6, 7] helps to explain the fine details of the concentration profiles observed for the interface between coexisting phases [130] or for the interfacial brush layer [261].

At the same time, understanding the basic features such as surface enrichment and related wetting phenomena (Sect. 3) still presents a significant challenge. Only very recent experimental observations of the entropy related mechanisms of the segregation (Sect. 3.1.2.2) and the enrichment-depletion duality (Sect. 3.1.2.4) have changed the classic viewpoint, relating the free surface segregation exclusively with the cohesive energy difference between blend components. Not completely resolved problems concern the interfacial forces driving the segregation – their nature, spatial range, temperature dependence, etc.

In most of experimental reports, the molecular properties of polymer blend components (such as choice of segments, size, architecture, or composition) have been varied while the "external" interface, confining the polymer blend, was fixed. It is, however, conceivable to tune the segregation properties by the proper modification of the "external" interface which bounds the polymer mixture. Initial works confirming this possibility [116, 163, 279] were published.

Most of the experiments reported so far have been performed for blends confined in a thin film geometry, but analyzed with a semi- infinite mixture approach. This is legitimate for thicker layers (Sect. 2.2.4) but it is not proper for thinner films, where both "external" interfaces of the film have to be considered simultaneously. Relatively little experimental work has been done (Sect. 3.2.2) in this field, in spite of its importance for potential applications. Therefore intensive research is expected, tracing finite size effects but also the phase domain morphologies with the advent of real three-dimensional profiling techniques.

Finally, it has been found that the deuterium "staining" of individual molecules, commonly used in condensed matter studies, in case of polymers can lead to serious consequences in bulk and surface thermodynamics. This was shown in this work by the phase separation of isotopic blends (Sect. 2.2.2), isotope swapping effect in blend miscibility (Sect. 2.2.3) and surface segregation (Sect. 3.1.2.5) as well as by the specific scaling law (Eq. 61) which governs the polymer brush conformation (Sect. 4.2.1).

Acknowledgements. The author dedicates this review to the memory of Zdzislaw Bober.
He thanks Prof. Jacob Klein and other colleagues from the Weizmann Institute of Science for their friendly and helpful assistance which enabled him to complete the ion beam experiments illustrating this work. He is also very indebted to J. Klein, L.J. Fetters, U. Steiner, T. Hashimoto, R. Brenn, K. Binder, R.A. Jerry, J. Genzer, G. Reiter, S. Kumar, and M. Stamm for all the collaboration and discussions benefiting this work.
The author is indebted to Mrs. G. Domoslawska for help in preparing the figures.
Finally, partial support of this work by the Reserve of the Rector of Jagellonian University is acknowledged.

References

1. Sanchez IC (1983) Ann Rev Mater Sci 13:387
2. Special issue of Physics World March (1995)
3. Special issue of MRS Bulletin January (1996)
4. de Gennes PG (1979) Scaling concepts in polymer physics. Cornell University, Ithaca
5. Strobl G (1996) The physics of polymers. Springer, Berlin Heidelberg New York
6. Binder K (1999) Adv Polymer Sci 138:1
7. Binder K (1995) Acta Polymer 46:204
8. de Gennes PG (1985) Rev Modern Phys 57:827
9. Koningsveld R, Kleintjens LA, Nies E (1987) Croatica Chemica Acta 60:53
10. Bates FS (1991) Science 251:898
11. Helfand E, Sapse AM (1975) J Chem Phys 62:1327
12. Klein J (1990) Science 250:640
13. de Gennes PG (1983) Physics Today 36:33
14. Steiner U, Klein J, Eiser E, Budkowski A, Fetters LJ (1992) Science 258:1126
15. Schmidt I, Binder K (1985) J Physique 46:1631
16. Budkowski A, Scheffold F, Klein J, Fetters LJ (1997) J Chem Phys 106:719
17. Krausch G (1995) Material Sci Eng R14:1
18. Budkowski A (1997) Metals Alloys Technologies 31:587
19. Chaturvedi UK, Steiner U, Zak O, Krausch G, Schatz G, Klein J (1990) Appl Phys Lett 56:1228
20. Payne RS, Clough AS, Murphy P, Mills PJ (1989) Nucl. Instrum. Meth B42:130
21. Kerle T, Scheffold F, Losch A, Steiner U, Schatz G, Klein J (1997) Acta Polymer 48:548
22. Mills PJ, Green PF, Palmstrom CJ, Mayer JW, Kramer EJ (1984) Appl Phys Lett 4:957
23. Genzer J, Rothman JB, Composto RJ (1994) Nucl Instrum Meth B 86:345
24. Coulon G, Russell TP, Deline VR, Green PF (1989) Macromolecules 22:2581
25. Withlow SJ, Wool RP (1989) Macromolecules 22:2648

26. Schwarz SA, Wilkens BJ, Pudensi MA, Rafailovich MH, Sokolov J, Zhao X, Zhao W, Zheng X, Russell TP, Jones RAL (1992) Mol Phys 76:937
27. Bernasik A, Rysz J, Budkowski A, Kowalski K, Camra J, Jedlinski J (1997) In: Olefjord I, Nyborg L, Bryggs D (eds) ECASIA '97, 7th European Conference on Applications of Surface and Interface Analysis. John Wiley, Chichester, p 775
28. Russell TP (1990) Materials Sci Reports 5:171
29. Kramer EJ (1991) Physica B 174:189
30. Herkt-Maetzky C, Schelten J (1983) Phys Rev Lett 51:896
31. Bates FS, Wignall GD, Koehler WC (1985) Phys Rev Lett 55:2425
32. Sawyer LC, Grubb DT (1987) Polymer Microscopy. Chapman and Hall, New York
33. Bielefeldt H, Hoersch I, Krausch G, Mlynek J, Marti O (1994) Appl Phys A59:103
34. Thomas EL (1986) Electron Microscopy. Wiley, New York
35. Albrecht TR, Dovek MM, Lang CA, Gruetter P, Quate CF, Kuan SNJ, Frank CW, Pease RFW (1988) J Appl Phys 64:1178
36. Flory PJ (1942) J Chem Phys 10:51
37. Huggins ML (1942) J Am Chem Soc 64:1712
38. Flory PJ (1953) Principles of polymer chemistry. Cornell University Press, Ithaca, New York
39. Sariban A, Binder K (1987) J Chem Phys 86:5859
40. Ginzburg VL (1960) Sov Phys Solid State 2:341
41. Schwahn D, Mortensen K, Yee-Madeira Y (1987) Phys Rev Lett 58:896
42. Taylor JK, Debenedetti PG, Graessley WW, Kumar SK (1996) Macromolecules 29:764
43. Bates FS, Muthukumar M, Wignall GD, Fetters LJ (1988) J Chem Phys 89:535
44. Pesci AI, Freed KF (1989) J Chem Phys 90:2017
45. Curro JG, Schweizer KS (1988) J Chem Phys 88:7242
46. Dudowicz I, Freed KF (1990) Macromolecules 23:1519
47. Dudowicz I, Freed KF, Lifschitz M (1994) Macromolecules 27:5387
48. Sanchez IC, Lacombe RH (1978) Macromolecules 11:1145
49. Binder K (1983) J Chem Phys 79:6387
50. Gehlsen MD, Rosedale JH, Bates FS, Wignall GD, Hansen L, Almdal K (1992) Phys Rev Lett 68:2452
51. Deutsch HP, Binder K (1992) Euro Phys Lett 17:697
52. Schweizer KS, Curro JG (1989) J Chem Phys 91:5059
53. Cahn JW, Hilliard JE (1958) J Chem Phys 28:258
54. Landau LD, Lifshitz EJ (1980) Statistical Physics, Oxford
55. de Gennes PG (1980) J Chem Phys 72:4756
56. Roe RJ (1986) Macromolecules 19:728
57. Jannik G, de Gennes PG (1986) J Chem Phys 48:2260
58. Tang H, Freed KF (1991) J Chem Phys 94:6307
59. Lifschitz M, Freed KF, Tang H (1995) J Chem Phys 103:3767
60. Flebbe T, Duenweg B, Binder K (1996) J Phys II (France) 6:667
61. Nakanishi H, Pincus P (1983) J Chem Phys 79:997
62. Shibayama M, Yang H, Stein RS, Han CC (1985) Macromolecules 18:2179
63. Sanchez IC (1989) Polymer 30:471
64. Cumming A, Wiltzius P, Bates FS (1990) Phys Rev Lett 65:863
65. Shen S, Torkelson JM (1992) Macromolecules 25:721
66. Huh W, Karasz FE (1992) Macromolecules 25:1057
67. Lohse DJ, Fetters LJ, Daoyle MJ, Wang H-C, Kow C (1993) Macromolecules 26:3444
68. Bates FS, Wignall GD (1986) Phys Rev Lett 57:1429
69. Bates FS, Fetters LJ, Wignall GD (1988) Macromolecules 21:1086
70. Schwahn D, Hahn K, Streib J, Springer T (1990) J Chem Phys 93:8383
71. Nicholson JC, Finerman TM, Crist B (1990) Polymer 31:2287
72. Balsara NP, Fetters LJ, Hadjichristidis N, Lohse DJ, Han CC, Graessley WW, Krishnamoorti R (1992) Macromolecules 25:6137

73. Londono JD, Narten AH, Wignall GD, Honnel KG, Hsieh T, Johnson TW, Bates FS (1994) Macromolecules 27:2864
74. Budkowski AS, Steiner U, Klein J, Schatz G (1992) EuroPhys Lett 18:705
75. Bruder F, Brenn R (1991) Macromolecules 24:5552
76. Chaturvedi UK, Steiner U, Zak O, Krausch G, Klein J (1989) Phys Rev Lett 63:616
77. Bruder F, Brenn R, Stuehn B, Strobl GR (1989) Macromolecules 22:4434
78. Klein J, Briscoe BJ (1975) Nature 257:387
79. Klein J, Briscoe BJ (1979) Proc Roy Soc (London) A 365:53
80. Green PF (1985) PhD Thesis, Cornell University, Ithaca, New York
81. Jones RAL (1987) PhD Thesis, Cambridge University
82. Kausch HH, Tirell M (1989) Annu Rev Mater Sci 19:341
83. de Gennes PG (1989) CR Acad Sci Paris, Serie II 308:13
84. Harden JL (1990) J Phys France 51:1777
85. Puri S, Binder K (1991) Phys Rev B 44:9735
86. Wang SQ, Shi Q (1993) Macromolecules 26:1091
87. Brochard F, Jouffroy J, Levinson P (1983) Macromolecules 16:1638
88. Steiner U (1993) PhD Thesis, Konstanz University
89. Steiner U, Krausch G, Schatz G, Klein J (1990) Phys Rev Lett 64:1119
90. Crank J (1975) The mathematics of diffusion. Oxford University Press, Oxford
91. Scheffold F, Eiser E, Budkowski A, Steiner U, Klein J, Fetters LJ (1996) J Chem Phys 104:8786
92. Budkowski A, Steiner U, Klein J (1992) J Chem Phys 97:229
93. Parry AO, Evans R (1992) Physica A 181:250
94. Straub W, Bruder F, Brenn R, Krausch G, Bielefeldt H, Kirsch A, Marti O, Mlynek J, Marko JF (1995) EuroPhys Lett 29:353
95. Bruder F (1992) PhD Thesis, Albert-Ludwigs Univ, Freiburg i. Br
96. Eiser E, Budkowski A, Steiner U, Klein J, Fetters LJ, Krishnamoorti R (1993) ACS PMSE Proc 69:176
97. Straub W (1996) PhD Thesis, Albert-Ludwigs Univ, Freiburg i. Br
98. Genzer J (1996) PhD Thesis, University of Pennsylvania
99. Genzer J, Composto RJ (1997) EuroPhys Lett 38:171
100. Cahn JW (1965) J Chem Phys 42:93
101. Pincus P (1981) J Chem Phys 75:1996
102. Hashimoto T (1988) Phase Transitions 12:47
103. Lifshitz IM, Slyoyzov VV (1961) J Chem Phys Solids 19:35
104. Ball RC, Essery RLH (1990) J Phys: Condens Matter 2:10,303
105. Marko JF (1993) Phys Rev E 48:2861
106. Puri S, Binder K (1994) Phys Rev E 49:5359
107. Budkowski A (unpublished)
108. Jones RAL, Norton LJ, Shull KR, Kramer E, Bates FS, Wiltzius P (1991) Phys RevLett 66:1326
109. Krausch G, Dai C-A, Kramer EJ, Bates FS (1993) Phys Rev Lett 71:3669
110. Krausch G, Mlynek J, Straub W, Brenn R, Marko JF (1994) EuroPhys Lett 28:323
111. Krausch G, Dai C-A, Kramer EJ, Marko JF, Bates FS (1993) Macromolecules 26:5566
112. Kim E, Krausch G, Kramer EJ, Osby JO (1994) Macromolecules 27:5927
113. Geoghegan M, Jones RAL, Clough AS (1995) J Chem Phys 103:2719
114. Bruder F, Brenn R (1992) Phys Rev Lett 69:624
115. Steiner U, Eiser E, Budkowski A, Fetters JL, Klein J (1994) Ber Bunsenges Phys Chem 98:366
116. Rysz J, Bernasik A, Ermer H, Budkowski A, Brenn R, Hashimoto T, Jedlinski J (1997) EuroPhys Lett 40:503
117. Steiner U, Klein J, Fetters LJ (1994) Phys Rev Lett 72:1498
118. Binder K, Nielaba P, Pereyra V (1997) Z Phys B 104:81
119. Krausch G, Kramer EJ, Rafailovich MH, Sokolov J (1994) Appl Phys Lett 64:2655

120. Scheffold F, Budkowski A, Steiner U, Eiser E, Klein J, Fetters LJ (1996) J Chem Phys 104:8795
121. Kerle T, Klein J, Binder K (1996) Phys Rev Lett 77:1318
122. Hildebrand JH, Scott RL (1950) The solubility of non-electrolytes. Reinhold
123. van Laar JJ, Lorentz R (1925) Z Anorg Allgem Chem 146:42
124. Coleman MM, Serman CJ, Bhagwar DE, Painter PC (1990) Polymer 31:1187
125. Lin J-L,Roe R-J (1987) Macromolecules 20:2168
126. Han CC, Bauer BJ, Clark JC, Muroga Y, Matsushita Y, Okada M, Tran-Cong Q, Chang I, Sanchez IC (1988) Polymer 29:2002
127. Dudowicz J, Freed KF (1991) Macromolecules 24:5112
128. Lifschitz M, Freed KF (1993) J Chem Phys 98:8994
129. Schubert DW, Stamm M (1996) EuroPhys Lett 35:419
130. Sferrazza M, Xiao C, Jones RAL, Bucknall DG, Webster J, Penfold (1997) Phys Rev Lett 78:3693
131. Majkrzak CF, Flecher GP (1990) MRS Bulletin November, p 65
132. Atkin EL, Kleintjens LA, Koningsveld R, Fetters LJ (1984) Makromol Chem 185:377
133. Sakurai S, Hasegawa H, Hashimoto T, Hargis IG, Aggarwal SL, Han CC (1990) Macromolecules 23:451
134. Rhee J, Crist B (1991) Macromolecules 24:5663
135. Larbi FBC, Leloup S, Halary JL, Monnerie L (1986) Polymer Communications 27:23
136. Rhee J, Crist B (1993) J Chem Phys 98:4174
137. Graessley WW, Krishnamoorti R, Balsara NP, Fetters LJ, Lohse DJ, Schultz DN, Sissano JA (1993) Macromolecules 26:1137
138. Krishnamoorti R, Graessley WW, Balsara NP, Lohse DJ (1994) J Chem Phys 100:3894
139. Graessley WW, Krishnamoorti R, Balsara NP, Fetters LJ, Lohse DJ, Schultz DN, Sissano JA (1994) Macromolecules 27:2574
140. Graessley WW, Krishnamoorti R, Balsara NP, Butera RJ, Fetters LJ, Lohse DJ, Schultz DN, Sissano JA (1994) Macromolecules 27:3896
141. Krishnamoorti R, Graessley WW, Balsara NP, Lohse DJ (1994) Macromolecules 27:3073
142. Graessley WW, Krishnamoorti R, Reichart GC, Balsara NP, Fetters LJ, Lohse DJ (1995) Macromolecules 28:1260
143. Budkowski A, Klein J, Eiser E, Steiner U, Fetters LJ (1993) Macromolecules 26:3858
144. ten Brinke G, Karasz FE, MacKnight WJ (1983) Macromolecules 16:1827
145. Budkowski A, Rysz J, Scheffold F, Klein J, Fetters LJ (1998) J Polym Sci: Part B: Polymer Physics 36:2691
146. Fredrickson GH, Liu AJ (1995) J Polym Sci: Part B: Polymer Physics 33:1203
147. Kohl PR, Seifert AM, Hellmann GP (1990) J Polymer Sci: Part B: Polym Physics 28:1309
148. Schwahn D, Janssen S, Springer T (1992) J Chem Phys 97:8775
149. Schwahn D, Meier G, Mortensen K, Janssen S (1994) J Phys II France 4:837
150. Rouault Y, Baschnagel J, Binder K (1995) J Statistical Physics 80:1009
151. Bates FS, Wignall GD (1986) Macromolecules 19:932
152. Young T (1805) Philos Trans R Soc London 95:65
153. Cahn JE (1977) J Chem Phys 66:3667
154. Moldover MR, Cahn JW (1980) Science 207:1073
155. Schmidt JW, Moldover MR (1983) J Chem Phys 79:379
156. Pohl DW, Goldburg WI (1982) Phys Rev Lett 48:1111
157. Ragil K, Meunier J, Broseta D, Indekeu JO, Bonn D (1996) Phys Rev Lett 77:1532
158. Dietrich S (1988) In: Domb C, Lebowitz J (eds.) Phase transitions and critical phenomena, 12. Academic Press, London, p 1
159. Jones RAL (1994) Polymer 35:2160
160. Steiner U, Klein J (1996) Phys Rev Lett 77:2526
161. Wang JS, Binder K (1991) J Chem Phys 94:8537
162. Genzer J, Composto RJ (1997) J Chem Phys 106:1257

163. Genzer J, Kramer EJ (1997) Phys Rev Lett 78:4946
164. Bhatia QS, Pan DH, Koberstein JT (1988) Macromolecules 21:2166
165. Jones RAL, Kramer EJ, Rafailovich M, Sokolov J, Schwarz SA (1989) Phys Rev Lett 62:280
166. Faldi A, Genzer J, Composto RJ, Dozier WD (1995) Phys Rev Lett 74:3388
167. Bruder F, Brenn R (1993) EuroPhys Lett 22:707
168. Ermer H, Straub W, Brenn R (1996) Verhand DPG (VI) 31:PO9.3
169. Jones RAL, Norton LJ, Kramer EJ, Composto RJ, Stein RS, Russell TP, Mansour A, Karim A, Felcher GP, Rafailovich MH, Sokolov J, Zhao X, Schwarz SA (1990) EuroPhys Lett 12:41
170. Zhao X, Zhao W, Sokolov J, Rafailovich MH, Schwarz SA, Wilkens BJ, Jones RAL, Kramer EJ (1991) Macromolecules 24:5991
171. Hariharan A, Kumar SK, Russell TP (1993) J Chem Phys 98:4163
172. Hariharan A, Kumar SK, Rafailovich MH, Sokolov J, Zheng X, Duong D, Russell TP (1993) J Chem Phys 99:656
173. Guckenbiehl B, Stamm M, Springer T (1994) Coll Surf A: Physicochem Eng Aspects 86:311
174. Genzer J, Faldi A, Oslanec R, Composto RJ (1996) Macromolecules 20:5438
175. Budkowski A, Rysz J, Scheffold F, Klein J (1998) EuroPhys Lett 43:404
176. Jerry RA, Dutta A (1994) J Coll Int Sci 167:287
177. Jerry RA, Nauman EB (1992) Phys Lett A 167:198
178. Kumar SK, Tang H, Szleifer I (1994) Molecular Physics 81:867
179. Rouault Y, Duenweg B, Baschnagel J, Binder K (1996) Polymer 37:297
180. Lipowski R, Huse DA (1986) Phys Rev Lett 52:353
181. Binder K, Frisch HL (1991) Z Phys B: Condensed Matter 84:403
182. Jones RAL, Kramer EJ (1990) Philos Mag B 62:129
183. Cohen S, Muthukumar M (1989) J Chem Phys 90:5749
184. Fredrickson GH, Donley JP (1992) J Chem Phys 97:8941
185. Jerry RA, Nauman EB (1992) J Coll Int Sci 154:122
186. Donley JP, Fredrickson GH (1995) J Polym Sci: Part B: Polymer Physics 33:1343
187. Carmesin I, Noolandi J (1989) Macromolecules 22:1689
188. Norton LJ, Kramer EJ, Bates FS, Gehlsen MD, Jones RAL, Karim A, Felcher GP, Kleb R (1995) Macromolecules 28:8621
189. Zink F, Kerle T, Klein J (1998) Macromolecules 31:417
190. Chen ZY, Noolandi J, Izzo D (1991) Phys Rev Lett 66:727
191. Jones RAL (1993) Phys Rev E 47:1437
192. Klein J, Scheffold F, Steiner U, Eiser E, Budkowski A, Fetters LJ (1997) In: Sperling LH (ed) Interfacial aspects of multicomponent polymer materials. Plenum, New York
193. Genzer J, Faldi A, Composto RJ (1994) Phys Rev E50:2373
194. Cifra P, Bruder F, Brenn R (1993) J Chem Phys 99:4121
195. Jones RAL, Kramer EJ (1993) Polymer 34:115
196. Hariharan A, Kumar SK, Russell TP (1991) Macromolecules 24:4909
197. Kumar SK, Russell TP (1991) Macromolecules 24:3816
198. Hariharan A, Kumar SK, Russell TP (1990) Macromolecules 23:3584
199. Walton DG, Mayes AM (1996) Phys Rev E54:2811
200. Mayes AM, Kumar SK (1997) MRS Bulletin January, p 43
201. Wu DT, Fredrickson GH (1996) Macromolecules 29:7919
202. Walton DG, Soo PP, Mayes AM, Allgor SJS, Fujii JT, Griffith LG, Ankner JF, Kaiser H, Johansson J, Smith GD, Barker JG, Satija SK (1997) Macromolecules 30:6947
203. Sikka M, Singh N, Karim A, Bates FS, Satija SK, Majkrzak CF (1993) Phys Rev Lett 70:307
204. Carigano MA, Szleifer I (1995) EuroPhys Lett 30:525
205. Wu DT, Fredrickson GH, Carton J-P (1996) J Chem Phys 104:6387
206. Donley JP, Wu DT, Fredrickson GH (1997) Macromolecules 30:2167

207. Yethiraj A, Kumar S, Hariharan A, Schweizer KS (1994) J Chem Phys 100:4691
208. Yethiraj A (1995) Phys Rev Lett 74:2018
209. Kumar S, Yethiraj A, Schweizer KS, Leermakers FAM (1995) J Chem Phys 103:10,332
210. Steiner U, Klein J, Eiser E, Budkowski A, Fetters LJ (1994) In: Rabin Y, Bruinsma R (eds) Soft order in physical systems. Plenum Press, New York
211. Steiner U, Eiser E, Budkowski A, Fetters JL, Klein J (1994) In: Teramoto A, Kobayashi M, Norisuje T (eds) ordering in macromolecular systems. Springer, Berlin Heidelberg New York
212. Scheffold F, Eiser E, Budkowski A, Steiner U, Klein J, Fetters LJ (1996) ACS PMSE Proc. 75 (Aug)
213. Schweizer KS, David EF, Singh C, Curro JG, Rajasekaran JJ (1995) Macromolecules 28:1528
214. van der Linden CC, Leermakers FAM, Fleer GJ (1996) Macromolecules 29:1172
215. Rysz J, Budkowski A, Bernasik A, Klein J, Kowalski K, Jedlinski J (1998) (in preparation)
216. Wu S (1982) Polymer interfaces and adhesion Marcel Dekker, New York
217. Guggenheim EA (1945) J Chem Phys 13:253
218. Balsara NP, Lohse DJ, Graessley WW, Krishnamoorti R (1994) J Chem Phys 100:3905
219. Tang H, Szleifer I, Kumar SK (1994) J Chem Phys 100:5367
220. Reich S, Cohen Y (1981) J Polymer Sci 19:1255
221. Nakanishi H, Fisher ME (1983) J Chem Phys 78:3279
222. Eggleton CD (1996) Physics Letters A 223:394
223. Müller M, Binder K (1998) Macromolecules (in press)
224. Werner A, Schmid F, Müller M, Binder K (1997) J Chem Phys 107:8175
225. Binder K, Müller M, Schmid F, Werner A (1997) J Computer-Aided Materials Design 4:137
226. Milner ST (1991) Science 251:905
227. Halperin A, Tirrell M, Lodge TP (1991) Advances in Polymer Science 100:31
228. Milner ST (1994) J Polym Sci B: Polymer Physics 32:2743
229. Alexander S (1977) J Phys (Paris) 38:983
230. de Gennes PG (1980) Macromolecules 13:1069
231. Leibler L (1988) Makromol Chem, Macromol Symp 16:1
232. Aubouy M, Raphaël E (1993) J Phys II France 3:443
233. Munch MR, Gast AP (1988) Macromolecules 21:1366
234. Zhulina EB, Borisov OV, Brombacher L (1991) Macromolecules 24:4679
235. Semenov AN (1992) Macromolecules 25:4967
236. Shull KR, Kramer EJ (1990) Macromolecules 23:4769
237. Shull KR (1991) J Chem Phys 94:5723
238. Shull KR (1993) Macromolecules 26:2346
239. Grest GS (1996) J Chem Phys 105:5532
240. Brown HR, Char K, Deline VR (1990) Macromolecules 23:3383
241. Budkowski A, Steiner U, Klein J, Fetters LJ (1992) EuroPhys Lett 20:499
242. Clarke CJ, Jones RAL, Edwards JL, Clough AS, Penfold J (1994) Polymer 35:4065
243. Clarke CJ, Jones RAL, Edwards JL, Shull KR, Penfold J (1995) Macromolecules 28:2042
244. Liu Y, Schwarz SA, Zhao W, Quinn J, Sokolov J, Rafailovich MH, Iyengar D, Kramer EJ, Dozier W, Fetters LJ, Dickman R (1995) EuroPhys Lett 32:211
245. Reiter G, Auroy P, Auvray L (1996) Centre de Recherches sur la Physico-Chimie des Surfaces Solides, CNRS, Mulhouse, France (preprint)
246. Green PF, Russell TP (1991) Macromolecules 24:2931
247. Russell TP, Anastasiadis SH, Menelle A (1991) Macromolecules 24:1575
248. Green PF, Russell TP (1992) Macromolecules 25:783
249. Zhao X, Zhao W, Zheng X, Rafailovich MH, Sokolov J, Schwarz SA, Pudensi MAA, Russell TP, Kumar SK, Fetters LJ (1992) Phys Rev Lett 69:776
250. Dai KH, Kramer EJ, Shull KR (1992) Macromolecules 25:220

251. Budkowski A, Klein J, Steiner U, Fetters LJ (1993) Macromolecules 26:2470
252. Dai KH, Washiyama J, Kramer EJ (1994) Macromolecules 27:4544
253. Dai KH, Kramer EJ (1994) J Polymer Sci B: Polymer Phys 32:1943
254. Budkowski A, Klein J, Fetters LJ (1995) Macromolecules 28:8571
255. Budkowski A, Klein J, Fetters LJ, Hashimoto T (1995) Macromolecules 28:8579
256. Shull KR, Kellock AJ, Deline VR, MacDonald SA (1992) J Chem Phys 97:2095
257. Ligoure C, Leibler L (1990) J Phys France 51:1313
258. Budkowski A, Losch A, Klein J (1995) Israel J Chem 35:55
259. Sanchez IC (1987) Encycl Polym Sci Technol 11:1
260. Shull KR, Kramer EJ, Hadziioannou G, Tang W (1990) Macromolecules 23:4780
261. Dai KH, Norton LJ, Kramer EJ (1994) Macromolecules 27:1949
262. Jones RAL, Norton LJ, Shull KR, Kramer EJ, Felcher GP, Karim A, Fetters LJ (1992) Macromolecules 25:2359
263. Dai KH, Kramer EJ (1994) Polymer 35:157
264. Nicolai T, Clarke CJ, Jones RAL, Penfold J (1994) Colloid Surf A86:155
265. Zhao X, Zhao W, Rafailovich MH, Sokolov J, Russell TP, Kumar SK, Schwarz SA, Wilkens BJ (1991) EuroPhys Lett 15:725
266. Losch A, Salomonovic R, Steiner U, Fetters LJ, Klein J (1995) J Polym Sci B: Polymer Physics 33:1821
267. Aubouy M, Fredrickson GH, Pincus P, Raphaël E (1995) Macromolecules 28:2979
268. Tanaka H, Hashimoto T (1991) Macromolecules 24:5398
269. Shull KR, Winey KI, Thomas EL, Kramer EJ (1991) Macromolecules 24:2748
270. Dan N, Tirrel M (1993) Macromolecules 26:6467
271. Milner ST (1992) Macromolecules 25:5847
272. Lai P-Y, Zhulina EB (1992) Macromolecules 25:5201
273. Laub CF, Koberstein JT (1994) Macromolecules 27:5016
274. Klein J, Kamiyama Y, Yoshizawa H, Israelachvili JN, Fetters LJ, Pincus P (1992) Macromolecules 25:2062
275. Kumacheva E, Klein J, Pincus P, Fetters LJ (1993) Macromolecules 26:6477
276. Perahia D, Wiesler DG, Satija SK, Fetters LJ, Sinha SK, Grest GS (1996) Physica B221:337
277. Char K, Brown HR, Deline VR (1993) Macromolecules 26:4164
278. Sung L, Han C (1995) J Polym Sci B: Polymer Phys 33:2405
279. Kellogg GJ, Walton DG, Mayes AM, Lambooy P, Russell TP, Gallagher PD, Satija SK (1996) Phys Rev Lett 76:2503

Editor: Prof. H. Höcker
Received: September 1998

Crystallization in Block Copolymers

I.W. Hamley

School of Chemistry, University of Leeds, Leeds LS2 9JT, UK
E-mail: I.W.Hamley@chem.leeds.ac.uk

Crystallization in block copolymers has a profound effect on their structure. This review article focusses on the morphology of semicrystalline block copolymers, and those containing two crystallizable blocks. The effect of crystallization on mechanical properties is briefly considered. The extent of chain folding upon crystallization is discussed, as is the orientation of crystal stems with respect to the microstructure. The effect of selective solvent on solution crystallization is also highlighted. Recent work on crystallization kinetics is summarized and finally the theories for crystallization in block copolymers are outlined.

Keywords: Block copolymers, Crystallization, Chain folding

Advances in Polymer Science, Vol.148
© Springer-Verlag Berlin Heidelberg 1999

List of Abbreviations

SANS	small-angle neutron scattering
SAXS	small-angle X-ray scattering
WAXS	wide-angle X-ray scattering
DSC	differential scanning calorimetry
TEM	transmission electron microscopy
LFRS	low frequency Raman spectroscopy
PE	poly(ethylene)
PEO	poly(ethylene oxide) [poly(oxyethylene)]
PCL	poly(ε-caprolactone)
PS	poly(styrene)
PEP	poly(ethylenpropylene)
PEE	poly(ethyl ethylene)
PB	poly(butadiene)
PVCH	poly(vinylcyclohexane)
PBO	poly(butylene oxide) [poly(oxybutylene)]
PPO	poly(propylene oxide) [poly(oxypropylene)]

1
Introduction

Crystallization of polymers is of great technological importance due to the mechanical properties imparted, which ultimately result from the change in molecular conformation. In semicrystalline block copolymers, the presence of a noncrystalline block enables modification of the mechanical and structural properties compared to a crystalline homopolymer, through introduction of a rubbery or glassy component. Crystallization in homopolymers leads to an extended conformation, or to *kinetically controlled* chain folding. In block copolymers, on the other hand, *equilibrium* chain folding can occur, the equilibrium number of folds being controlled by the size of the second, noncrystallizable block.

The most important crystallizable block copolymers are those containing poly(ethylene) or poly(ethylene oxide) (PEO) [systematic name poly(oxyethylene)]. Poly(ethylene) (PE) in block copolymers is prepared by anionic polymerization of poly(1,4-butadiene) (1,4-PB) followed by hydrogenation and has a melting point in the range 100–110 °C. This synthetic method leads to ethyl branches in the copolymer, with on average 2–3 branches per 100 repeats. These branches induce lengths for folded chains which are set by the branch density and not by the thermodynamics of crystallization. The melting temperature of PEO in block copolymers is generally lower than that of PEO homopolymer (melting temperature T_m=76 °C for high molecular weight samples). In contrast to PE, prepared by hydrogenation of 1,4-PB, there is no chain branching in these copolymers and the fold length depends on the crystallization procedure. Molecules with 1,2,3 ... folds can be obtained by varying the crystallization protocol (quench depth, annealing time etc.).

Crystallization has been investigated for other block copolymers, in particular those containing poly(ε-caprolactone) (T_m=57 °C). Experimental results for these materials are also summarized here. The morphology in block copolymers where both blocks are crystallizable is also discussed. However, except in section 7, the remainder of this review concerns semicrystalline block copolymers. Elements of this review have appeared in an extended form elsewhere [1].

2
Mechanical Properties

Crystallization in semicrystalline block copolymers can have a dramatic impact on mechanical properties, and hence is important to end use. There have been numerous studies, in particular of PE-containing block copolymers including industrial materials. A comprehensive overview of this work is outside the scope of this review. Instead, an illustrative example serves to illustrate the essential physics.

The mechanical and thermal properties of a range of poly(ethylene)/poly(ethylene propylene) (PE/PEP) copolymers with different architectures have been compared [2]. The tensile stress-strain properties of PE-PEP-PE and PEP-PE-PEP triblocks and a PE-PEP diblock are similar to each other at high PE content. This is because the mechanical properties are determined predominantly by the behaviour of the more continuous PE phase. For lower PE contents there are major differences in the mechanical properties of polymers with different architectures, that form a cubic-packed sphere phase. PE-PEP-PE triblocks were found to be thermoplastic elastomers, whereas PEP-PE-PEP triblocks behaved like particulate filled rubber. The difference was proposed to result from bridging of PE domains across spheres in PE-PEP-PE triblocks, which acted as physical crosslinks due to anchorage of the PE blocks in the semicrystalline domains. No such arrangement is possible for the PEP-PE-PEP or PE-PEP copolymers [2].

3
Morphology

Structural changes in block copolymers containing a crystallizable component, in particular chain-folding, resulting from crystallization compete with those occurring due to microphase separation. Experiments suggest that the final morphology after crystallization depends on whether the sample is cooled from a microphase-separated melt or crystallizes from a homogeneous melt or solution [3–6]. This path dependence is a general feature of crystallization in block copolymers. Solvents which are selective for the amorphous block can lead to non-equilibrium morphologies because the crystallizable block can precipitate from solution and crystallize. For example, when a poly(ethylene)-poly(styrene) (PE-PS) diblock was solvent cast from toluene above T_m(PE), crystallization occurred within microphase-separated PE domains whereas an irregular structure resulted when solvent was removed below T_m [5].

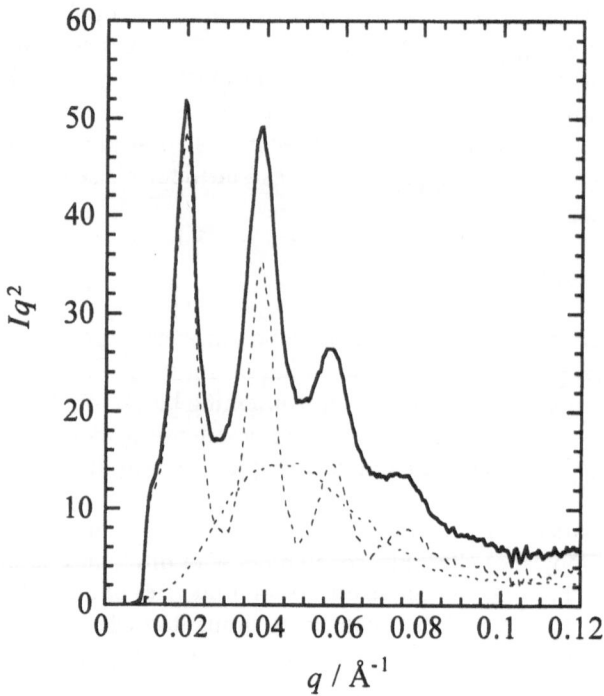

Fig. 1. Plot showing that the SAXS intensity from a PE-PEE diblock (M_n=23 kg mol^{-1}, f_{PE}= 0.49) at a temperature below the PE crystallization temperature can be represented as the sum of a broad peak from amorphous and crystalline PE (*dashed curve with one peak*) plus the multiple Bragg peak scattering from a lamellar structure (*dashed curve with four peaks*) [11]

Small-angle X-ray scattering (SAXS) is widely used to investigate the morphology of semicrystalline block copolymers at the level of the microphase separated structure, whilst wide-angle X-ray scattering (WAXS) is used to probe structure at the length scale of the crystal unit cell. Whereas X-ray scattering probes a global average of the structure, the local morphology can be imaged via transmission electron microscopy (TEM) where sectioned samples are selectively stained, highlighting features such as lamellae. For example, staining of PE-containing block copolymers is possible using RuO_4, which produces contrast due to reduced diffusivity in the crystalline domains [7].

The SAXS profiles from crystalline PE-containing diblocks corresponds to the sum of scattering from block copolymer lamellae plus a broad peak arising from semicrystalline PE, as shown in Fig. 1. As usual in SAXS patterns, the abscissa is wave vector $q=4\pi\sin\theta/\lambda$ where 2θ is the scattering angle and λ is the wavelength. Further information on the structure can be deduced from SAXS via the scattering density correlation function, which enables the thickness of the crystalline region to be determined from the extrapolated linear region of the scattering density correlation function [8].

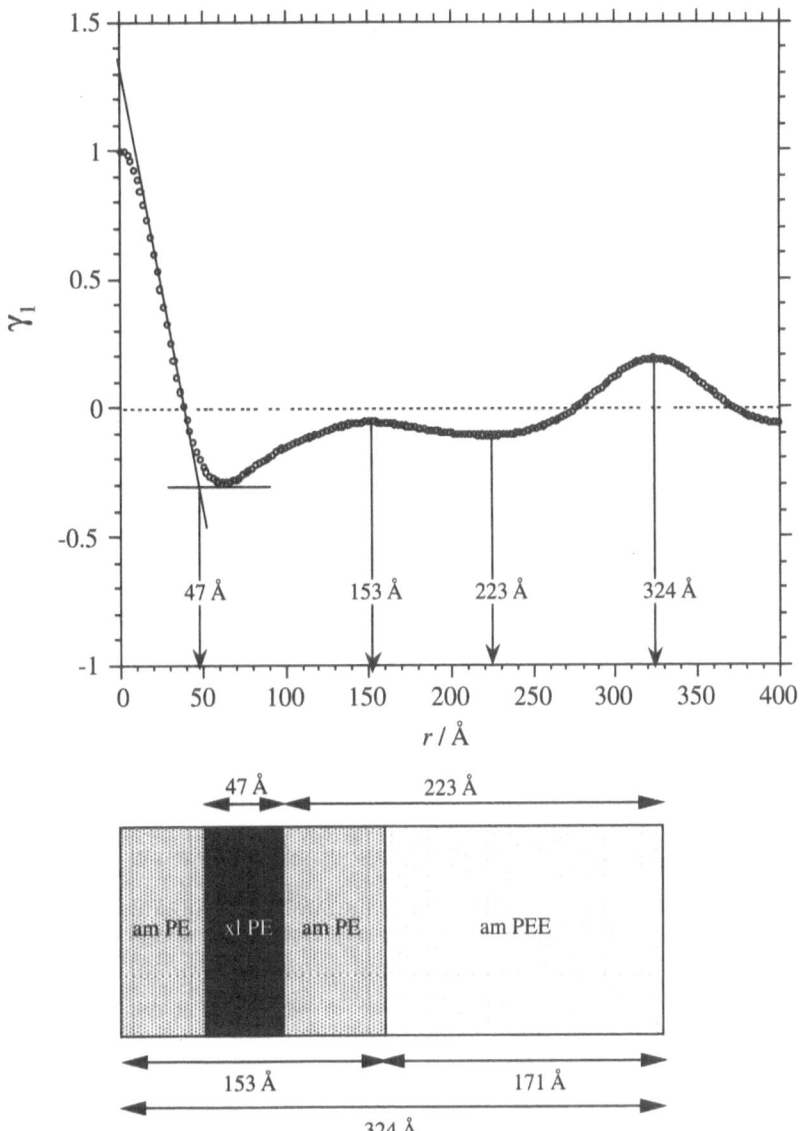

Fig. 2. Cosine-transform correlation function computed from the SAXS data for a PE-PEE diblock (M_n=23 kg mol^{-1}, f_{PE}=0.49) with an interpretation of the distances based on extrapolation of the small r slope and subsequent maxima and minima [11]

This method was used to determine the structure of a number of PE-containing diblocks. An example of a correlation function calculated for a PE-poly(ethyl ethylene) diblock, together with its interpretation in terms of amorphous and crystalline layer thicknesses is shown in Fig. 2. The PE crystal thickness in these polymers is set by the density of ethyl branches in the PE, which was synthesized by hydrogenation of poly(1,4-butadiene) [9]. In contrast to the constant PE crystal thickness, the lamellar domain spacing of PE homopolymers and PE-containing diblocks was found to decrease with increasing quench depth below T_m(PE), reflecting an increased degree of crystallinity.

The degree of crystallinity can be obtained from differential scanning calorimetry (DSC), SAXS or WAXS experiments. From DSC the degree of crystallinity $X=\Delta H_{exp}/\Delta H^0_{fus}w$, where ΔH_{exp} is the experimental melting enthalpy, w is the weight fraction of crystallizable block, and ΔH^0_{fus} is the theoretical heat of fusion for the crystalline homopolymer. The degree of crystallinity can also be obtained from SAXS data as l_c/d', where l_c is the crystalline region thickness and d' is the total crystallizable component thickness ($d'=df_c$, where f_c is the volume fraction of crystallizable component), both extracted from a correlation function analysis. The *relative* degree of crystallinity can be tracked using the small-angle scattering invariant. This is the integrated (q^2-weighted) intensity and provides a measure of the total small-angle scattering from a material, independent of the size or shape of structural inhomogeneities [10]. The *relative* degree of crystallinity can also be estimated from the WAXS data from the ratio of the integrated intensity of the crystal peak to that of the total amorphous and crystalline scattering [10]. For PE, the amorphous scattering below the (110) peak is relatively insensitive to the degree of crystallinity, so the integrated area of the (110) reflection compared to the broad amorphous halo is directly proportional to X_{PE}. However, the *absolute* degree of crystallinity cannot be determined in this way [11].

SAXS has been employed to examine the crystallization of poly(ε-caprolactone) in poly(ε-caprolactone)-poly(butadiene) (PCL-PB) diblock copolymers [12]. The melting temperature of PCL in the diblocks was up to 15 °C lower than the homopolymer value, T_m=57 °C. A sample with an order-disorder temperature, $T_{ODT}<T_m$ quenched directly from the disordered melt did not show the sharp diffraction peaks characterizing a microphase-separated phase, i.e. crystallization proceeded directly from the homogeneous state. However, for two samples with $T_{ODT} \sim T_m$ quenched from the homogeneous melt, microphase separation preceded crystallization but the ordered melt morphology was destroyed by crystallization. An important factor in these observations is the proximity of T_m to T_{ODT} [13]. It was suggested that if crystallization occurs just below T_{ODT}, the energy barrier to disruption of the microphase-separated morphology can readily be overcome. An additional factor is the incompatibility of the diblock [13]. The energy barrier between microphase-separated structure and crystal is expected to become larger with increasing molecular weight and/or interaction parameter χ.

The structure of PE-PEE, PE-PEP and PE-PVCH diblocks upon crystallization from the ordered melt has been elucidated using simultaneous SAXS, WAXS and DSC [9,11]. Here, PEE and PEP are rubbery components, whereas PVCH=poly(vinycyclohexane) is glassy below a glass transition temperature $T_g \sim 140$ °C. It was observed that the melt morphologies were destroyed due to PE chain folding

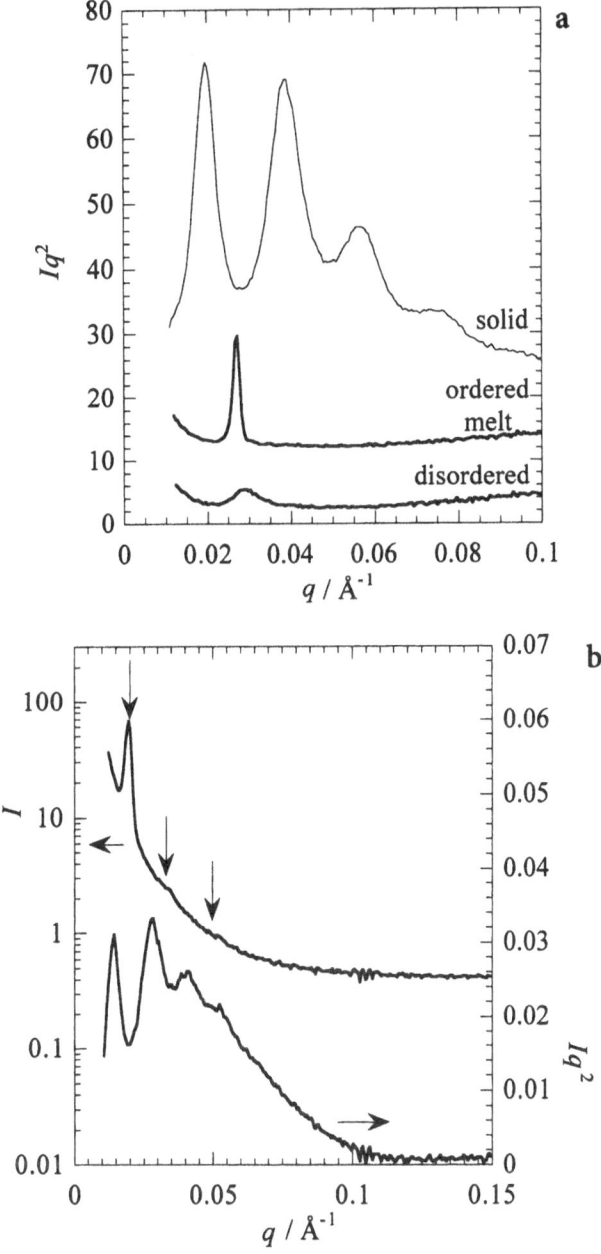

Fig. 3. SAXS intensity vs. scattering vector magnitude for PE-PEE diblocks with **a** M_n=120 kgmol^{-1}, f_{PE}=0.50 (*top two profiles* at 100 °C, *bottom profile* at 140 °C), **b** M_n=47 kg mol^{-1}, f_{PE}=0.25 [11]

upon crystallization. Whether the initial structure was an ordered lamellar or hexagonal-packed cylinder melt phase, after crystallization a lamellar phase distinct from the melt was always formed. This is illustrated in Fig. 3 which shows SAXS profiles for symmetric and asymmetric PE-PEE diblocks before and after crystallization. In contrast, crystallization from the strongly segregated cylindrical melt of PE-poly(3-methyl-1-butene) diblocks occurred within the preexisting cylindrical structure [14–16]. For a more weakly segregated diblock of the

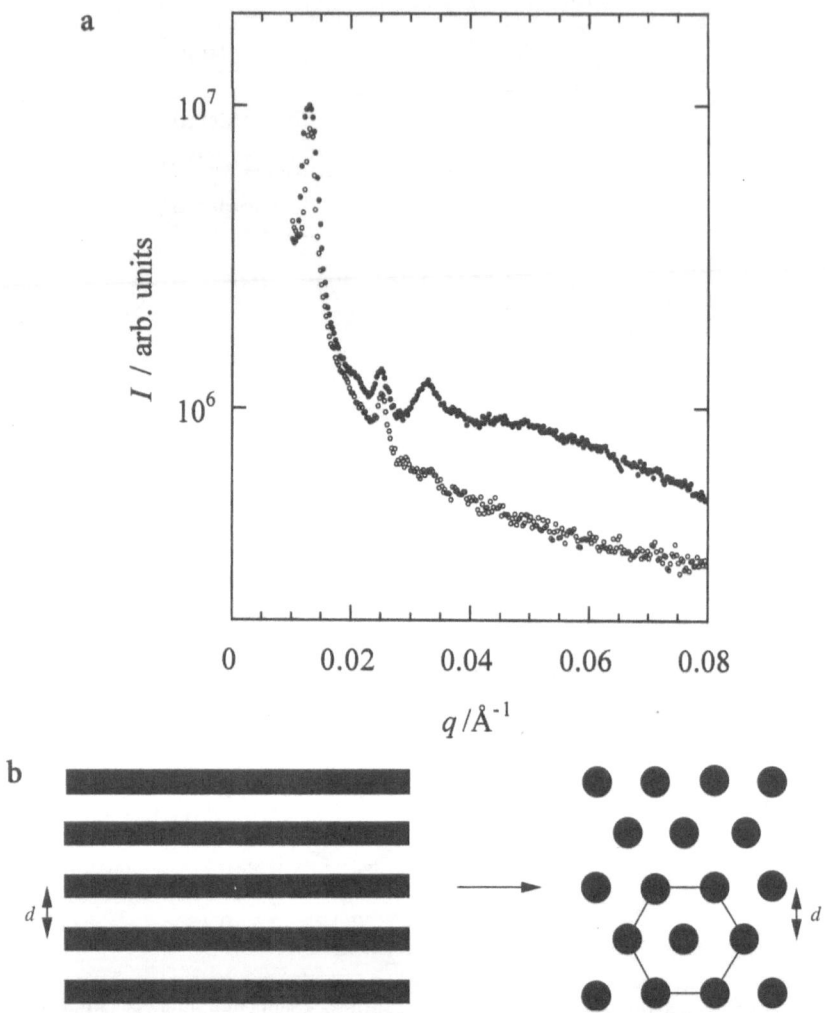

Fig. 4. a SAXS intensity profiles along the lamellar normal for a PE-PEE diblock (M_n=81 kg mol^{-1}, f_{PE}=0.35) [17]; **b** schematic of the epitaxial growth of a hexagonal-packed cylinder melt structure from a lamellar solid structure. The direction of incidence of X-rays corresponding to (a) is along the direction of the arrow

same type, crystallization within cylinders occurred upon fast cooling from the ordered melt, whereas slower cooling led to a lamellar structure [15,16]. These observations suggest that the structure with crystalline rods is metastable.

Epitaxial transitions were noted on melting asymmetric PE-containing diblocks with compositions f_{PE}=0.35 and 0.46 [17]. A lamellar structure melted epitaxially (i.e. the domain spacing and orientation were maintained across the transition) to a hexagonal-packed cylinder structure in the f_{PE}=0.35 sample. The low temperature lamellar structure implies that crystallization does not occur in cylinders. The epitaxy is illustrated in Fig. 4, which shows SAXS patterns in the solid and melt states, with a schematic of the epitaxial melting process [17,18]. The same epitaxial transition was observed for a PEO-PBO diblock [19].

4
Domain Spacing Scaling

Self-consistent field theory leads to the prediction that the domain spacing, d, for a lamellar semicrystalline diblock, scales as [20]:

$$d \sim NN_a^{-5/12} \tag{1}$$

where N is the total degree of polymerization and N_a is the degree of polymerization of the amorphous block. Cohen's group has reported results from SAXS/WAXS experiments on PE-PEE diblocks in accord with this prediction [21]. Rangarajan et al. reported $d \sim NN_a^{-0.45}$, for a series of PE-PEP diblocks also in good agreement with this scaling. However, they noted that theory does not account for the absolute values of d, because it only considers uniformly folded

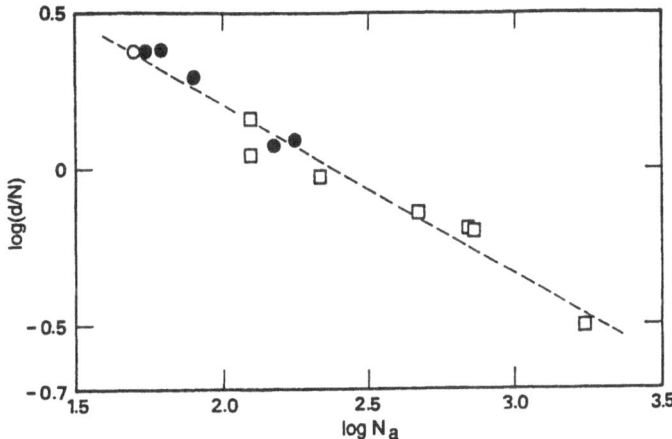

Fig. 5. Scaling of lamellar domain spacing reduced by N with the degree of polymerization of the amorphous block, N_a, for various semicrystalline diblocks [25]. (□) Data for PE-PEP diblocks from [22], (○) data for a PCL-PDMS-PCL triblock from [60], (●) data for PCL-PB diblocks from [25]

chains, which is unrealistic due to irregular folds induced by the presence of ethyl branches in these PE-containing diblocks [22]. The scaling of the lamellar thickness with degree of polymerization of the crystalline and amorphous blocks has also been investigated for PEO-poly(*tert*-butylmethacrylate) diblocks [23]. The experimental data for the scaling of the crystal layer thickness with degree of polymerization of the amorphous block were in agreement within the experimental errors with the theories of DiMarzio [24] and Whitmore and Noolandi [20] [Eq. (1)], being closer to the latter. However, the scaling with total degree of polymerization was not assessed. The lamellar thickness of a series of PCL-PB diblocks has been measured using SAXS [25]. It was found that the lamellar thickness for the amorphous layer increases more slowly with N than that for melt phases. The overall domain spacing was found to scale as $d \sim N N_a^{-\beta}$, the exponent $\beta = 0.54 \pm 0.04$ being in close agreement with the predictions of the Whitmore–Noolandi theory ($\beta = 5/12$).

Results from experiments to determine the domain spacing scaling are collated in Fig. 5. The scaling of domain spacing with N is stronger than the 2/3 law found for strongly segregated block copolymer melts, although a weaker function of N_a.

5
Chain Folding

Wide-angle X-ray scattering shows that PE in block copolymers crystallizes in the orthorhombic form that is also the crystalline structure of homopolymer PE at normal densities. PEO crystallizes in its usual helical form (alternating right- and left-hand distorted 7/2 helices), whilst PCL crystallizes as a planar zig-zag. For PE prepared by hydrogenation of 1,4-PB, chain folding is controlled by ethyl branch density. In contrast, in PEO- or PCL-containing block copolymers whether chains crystallize in an extended or crystalline conformation depends, in equilibrium, on the lengths of both crystalline and non-crystalline blocks. However, kinetically induced folding can also occur.

Evidence for the effect of chain ends on interfacial energy is provided by a number of experiments. The melting points of short PEO-containing diblocks [26] and triblocks [27,28], $T_m = 47–51$ °C, is low compared to perfectly crystalline PEO ($T_m = 76$ °C). This is due to the positive free energy of formation for the block copolymers of the amorphous layer from the melt, and of the crystalline/amorphous interface. The end interfacial theory can be analyzed in terms of the theories for melting points of low molecular weight polymers [29,30].

Lamellar spacings and melting points for a series of PPO-PEO-PPO triblocks have been reported [PPO=poly(propylene oxide)=poly(oxypropylene)] [27]. These properties were compared for $PPO_n PEO_m PPO_n$ triblocks with m=48 and different values of n in the range 0 to 7. The properties were interpreted using the Flory–Vrij model [30]. For copolymers with n≤2, extended chain crystals were observed whereas longer chains were once and/or twice folded, the melting point decreasing for crystals containing more highly folded chains. Chain folding in semicrystalline block copolymers can lead to more than one melting transition, and these can be assigned to crystals with different numbers of folds. An example is provided by results for $PPO_nPEO_mPPO_m$ triblocks with varying n and m [28]. The higher melting transition corresponds to that of the least folded crystal (i.e. extended or once-folded chain). For a given copolymer, the end interfacial energy σ_e is lower for more highly folded PEO chains [28]. For a given crystal type (e. g. once-folded chain), the end interfacial energy σ_e is higher for longer end blocks.

A useful technique for studying chain folding that has been applied to PEO-containing diblocks is low frequency Raman spectroscopy (LFRS). Low frequency Raman transitions can be assigned to longitudinal acoustic modes (LAM) of either unfolded or folded chains. The crystal structures of $PPO_nPEO_mPEO_n$ triblocks with m=39 or 75 and end block lengths in the range n=1–13, have been investigated using SAXS and DSC and LFRS [31]. For these triblocks there was evidence for polymers with n>1 of a structure comprising lamellae tilted with respect to the lamellar plane by up to 40–50°, whether the PEO blocks were folded or not. Regardless of the possibility of tilting, in general short chains adopt an extended conformation, whereas longer chains fold. Similar conclusions have been reached for poly(ethylene oxide) capped with alkoxy end groups, i.e. in highly asymmetric block copolymers [32–36]. The enthalpy of fusion also provides an indicator of whether chains are folded or extended. For example, for a series of end-capped PEO polymers, the enthalpy takes values ΔH^0_{fus}= (170±5) J g^{-1} for extended chain crystals and ΔH^0_{fus}=(145±5) J g^{-1} for folded chain crystals, irrespective of the end group [33].

In PEO-PBO diblocks with short blocks, unfolded PEO blocks crystallize into lamellar crystals, and the PBO blocks are also constrained to an unfolded (trans-planar) conformation. This was demonstrated experimentally by a combination of SAXS and LFRS used to study the series of block copolymers PEO_mPBO_n, with m approximately constant (m=26–32 units) and n in the range 3–30 units [37]. Copolymers with n≤20 form lamellae in which the whole chain is in extended conformation. As the PBO block length is increased the lamellae become increasingly disordered. However, if the PBO block length is increased to 30 units, the unfolded conformation of the PEO_mPBO_n copolymers becomes unstable with respect to a folded conformation [37]. Folding is a direct result of the difference in cross-sectional areas of the two chains coupled with the need to fill space at approximately normal density.

The crystallization of PEO-PBO diblock copolymers with PEO block lengths of 70 or more has been studied using simultaneous SAXS and WAXS [19]. The copolymers crystallized from melts (disordered, lamellar or hexagonal) with folded PEO blocks. Rapid crystallization of lamellar melts was accompanied by a change of length scale and led to kinetically determined multiply folded PEO blocks and to slight stretching of the PBO blocks from their melt conformation. This is apparent in Fig. 6 for a PEO-PBO diblock crystallized under different conditions. Quenching the polymers from a hexagonal melt phase into the solid state resulted in no change in length scale and led to structures with multiply

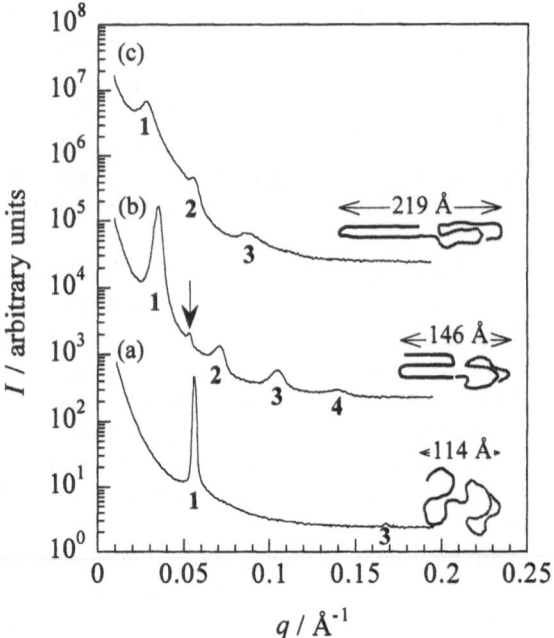

Fig. 6. SAXS patterns for $PEO_{76}PBO_{38}$ showing (*a*) the ordered melt structure (*T*=90 °C), (*b*) a metastable structure at *T*=42 °C, and (*c*) the equilibrium once-folded structure grown at *T*=50 °C by a self-seeding process [19]. Numbers indicate the order of reflection from a lamellar structure and the arrow indicates the position of the peak in the ordered melt. The calculated repeat lengths for possible molecular conformations are indicated

folded PEO blocks and PBO blocks unstretched from their melt conformation. That no change in length scale was associated with the crystallization process suggests an epitaxial relationship between the melt and crystalline structure, as discussed in the preceding section. The metastable folded structures that were formed during rapid crystallization did not unfold on heating but could be self-seeded to grow equilibrium structures with fewer folds. These folded states are controlled by a balance of contributions to the Gibbs energy from PEO block folding (positive contribution) and PBO block relaxation (negative contribution). The kinetically induced folded structures of PEO homopolymers have much larger stem lengths than those of the equilibrium folded structures of the diblocks. This emphasizes the importance of the competition between the two low Gibbs energy conformations in the block copolymers, i.e. unfolded PEO blocks and unstretched PBO blocks [19].

A more extensive study of crystallization in PEO-PBO diblocks with both ordered (hexagonal, lamellar and gyroid) and disordered melts has been published [38]. The effect on the melting point of crystallization under different conditions and lamellar spacings was investigated using DSC and SAXS, respec-

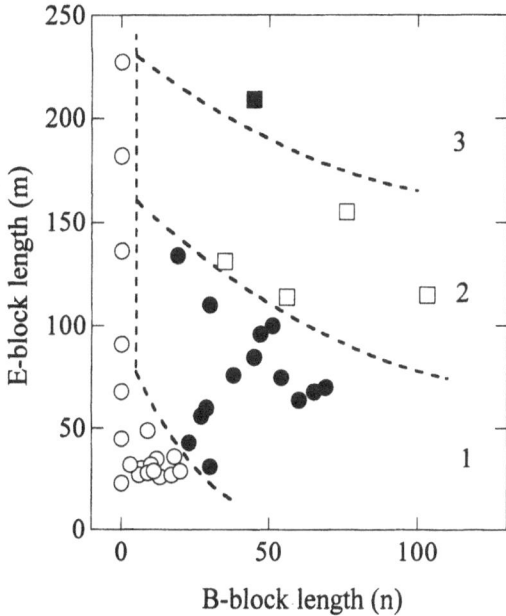

Fig. 7. The effect of PEO and PBO block lengths on the extent of chain folding of the PEO block in the equilibrium states of crystalline PEO_mPBO_n copolymers: (O) unfolded, (●) one fold, (□) two folds, (■) three folds [38]. The *dashed curves* approximately delineate the four regions. The data points at $m=0$ indicate the equilibrium state (unfolded) of low molar mass poly(oxyethylene)s

tively. Samples were cooled or quenched from above the melt temperature, annealed at a temperature below T_m or were examined as removed from storage without further treatment. Self-seeding was achieved by heating to the melting point, quenching to T_m −5 °C, holding for 12–24 h and finally quenching. It was found that the lamellae in all crystals formed by quenching could be thickened by annealing. However, the d spacings obtained from SAXS experiments after a period of annealing did not match the larger spacings obtained by self-seeded crystallization at high temperature. A model of helical PEO chains and oriented PBO chains was best able to account for the folding of both blocks. In this model the orientation of PBO chains emerging from the crystal end surface is maintained, so as to form an array of PBO blocks similar to a smectic liquid crystal. In contrast, in a 'normal density' model both blocks would have their normal densities, but this is inappropriate for short (extended) PBO blocks. Both PEO and PBO blocks can fold; however, due to the smaller cross section of PEO, only conformations with more folds in the PEO block are usually accessed. Denoting the number of folds by (f,g) where f is the number of folds in the PEO block and g is the number of PBO folds it was found that conformations with f=g are disfavoured due to the mismatch in block cross-sectional areas and densities. These results have been summarized in a diagram relating the extent of folding in the PEO block in the equilibrium state to the PEO and PBO block lengths in the copolymer (Fig. 7) [38].

6
Orientation of Folded Chains

The orientation of crystalline stems with respect to the interface of the microstructure in block copolymers depends on both morphology and the speed of chain diffusion, which is controlled by block copolymer molecular weight and the crystallization protocol (i.e. cooling rate). In contrast to homopolymers, where folding of chains occurs such that stems are always perpendicular to the lamellar interface, a parallel orientation was observed for block copolymers crystallized from a lamellar melt phase; perpendicular folding was observed in a cylindrical microstructure. Both orientations are shown in Fig. 8. Chain orientation can be probed via combined SAXS and WAXS on specimens oriented by shear or compression. In PE, for example, the orientation of (110) and (200) WAXS reflections with respect to Bragg peaks from the microstructure in the SAXS pattern enables the unit cell orientation to be deduced. Since PE stems are known to be oriented along the c axis, the chain orientation with respect to the microstructure can be determined.

The orientation of PE stems with respect to the lamellar microstructure has been determined for an asymmetric PE-PEP-PE triblock using SAXS and WAXS

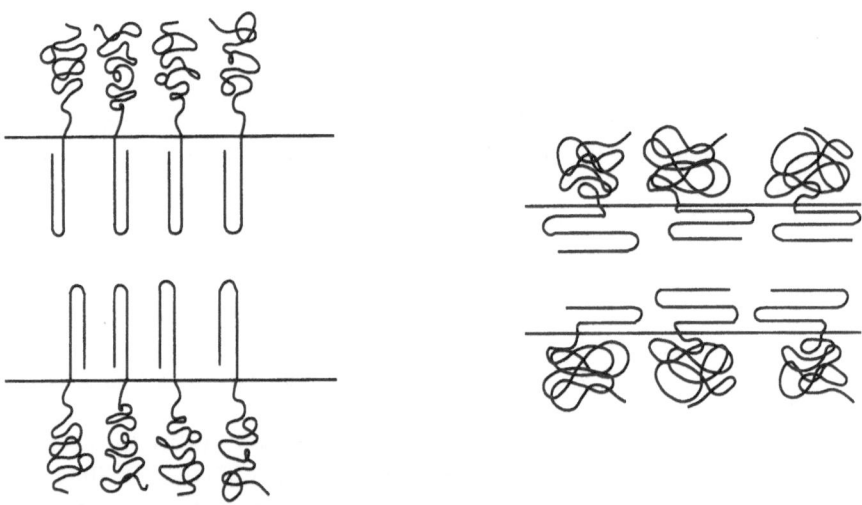

Perpendicular folding   Parallel folding

Fig. 8. Schematic of perpendicular and parallel chain folding in semicrystalline block copolymers

on samples oriented by drawing [4]. For a sample crystallized from a hexagonal melt phase, four-point SAXS patterns developed at relatively low draw ratios. These are often observed for the hexagonal phase of triblocks subject to large uniaxial strains [39]. At a higher draw ratio, the four-point pattern evolved into a pair of Bragg peaks normal to the draw direction, characteristic of a layered structure. At this draw ratio the (110) and (200) WAXS reflections from PE oriented in the same direction as the peaks in the SAXS arising from the lamellar structure. Following relaxation, the (110) peaks shifted away from the normal to the draw direction to an azimuthal angle $\phi=56°$, to give a pattern characteristic of the projection along the a axis of oriented body-centred orthorhombic PE. Thus the PE stems were parallel to the lamellae. In contrast, for a solution crystallized sample a WAXS pattern characteristic of orientation of the PE unit cell b axis along the draw direction was observed, although most of the anisotropy in the WAXS was lost following relaxation. The corresponding SAXS patterns did not contain sharp peaks characteristic of highly ordered samples, only diffuse scattering features. At very high draw ratios the WAXS patterns contained reflections from monoclinic PE as well as the usual orthorhombic form [4].

Chain orientation parallel to the lamellar interface was also reported for PE-PEE diblocks crystallized from the ordered melt [40]. In subsequent work, the chain orientation in a PE-PS diblock was investigated (here the ethyl branch density in PE was reduced to essentially zero) [41]. The PE stems were again par-

(a)   (b)

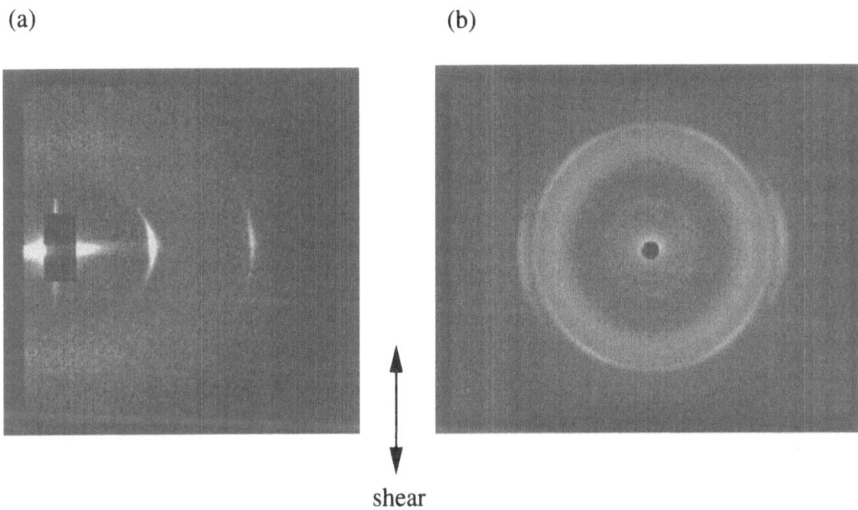

shear

Fig. 9. Scattering patterns obtained for a PE-PVCH diblock (M_n=15 kg mol^{-1}, f_{PE}=0.52) at room temperature [17]. **a** SAXS pattern. The sharp first-order lamellar reflection is at q^*= 0.035 Å^{-1}; **b** WAXS pattern. Four 110 reflections are apparent at $2\theta(CuK_\alpha)$=21°, and two equatorial 200 reflections at $2\theta(CuK_\alpha)$=24°. The X-ray beam was incident perpendicular to the shear direction and to the lamellar normal (i.e. along the perpendicular direction in Fig. 10)

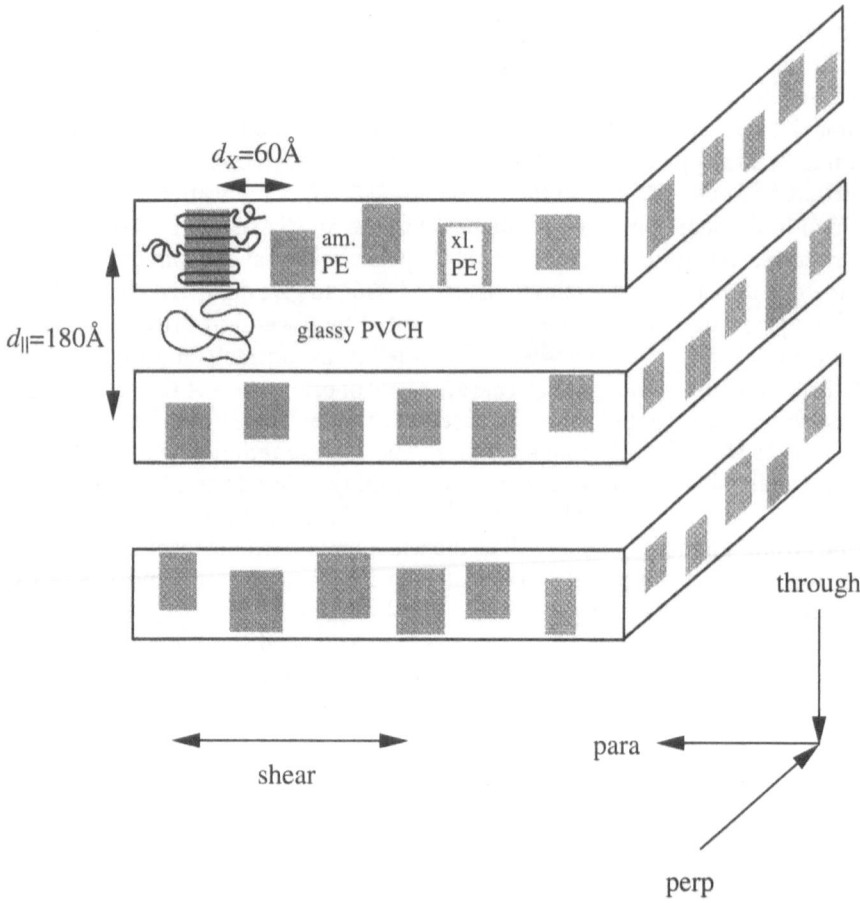

Fig. 10. Model for the lamellar organization in semicrystalline PE-containing diblocks crystallized from the ordered melt [17]. The PE chains are folded with stems parallel to the lamellar interface. The convention for labelling of the axis system with respect to the shear direction is also indicated

allel to the lamellar interface, i.e. the a axis of the unit cell was oriented normal to the lamellae, with the b and c axes (and thus stems) lying with random orientation in the lamellar plane. Orientation in a series of PE-PEP diblocks aligned by compression in a channel die was also studied using SAXS/WAXS [42]. The lamellar alignment was varied with respect to the shear plane (by changing temperature), being either perpendicular or parallel. The PE chains crystallized parallel to the lamellae in either alignment [42].

The state of the non-crystalline block can have a profound effect on crystallization in semicrystalline diblocks. Crystallization in PE-PEE, PE-PEP and PE-PVCH diblock copolymers has been investigated for samples oriented by shear-

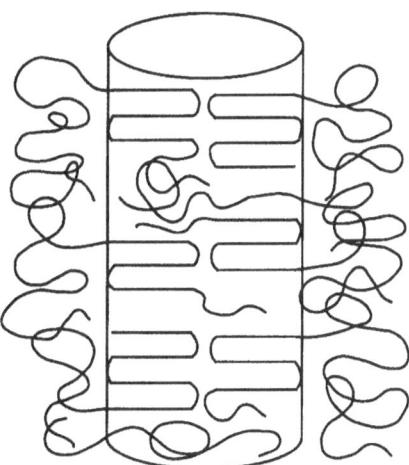

Fig. 11. Perpendicular crystalline stem orientation with respect to cylinders

ing in the ordered melt and quenching to room temperature [17,18]. For a PE-PVCH sample this led to crystallization within a lamellar phase with glassy PVCH lamellae [recall that T_g(PVCH)>T_m(PE)]. For symmetric diblocks with either a rubbery or glassy amorphous block, the orientation of the crystallized PE stems was parallel to the lamellar interface, as noted for other block copolymers [4,40–42]. For the PE-PVCH diblock, diffuse scattering was observed normal to the shear direction in the SAXS pattern (Fig. 9). This is consistent with lateral correlations between PE crystallites within the layers of semicrystalline PE, as sketched in Fig. 10. In contrast, in all the samples containing an amorphous component, PE crystallization occurred with no lateral positional correlations of crystallites. The diffuse scattering for the PE-PVCH diblock was modelled using a Markov lattice. In particular, a lattice was constructed from a stack of one-dimensional Markov sequences, where the probability of a site crystallizing depends only on the occupancy of the previous site in that sequence. Diffuse scattering bars normal to the deformation axis were observed earlier in SAXS patterns obtained by drawing a melt-crystallized PE-PEP-PE triblock [4].

Although Rangarajan et al. did not study oriented samples, they suggested a model for their PE-poly(ethylene-*alt*-propylene) diblocks, crystallized from the *homogeneous* melt in which PE chains are folded perpendicular to the lamellar interface [22]. This differs from the orientation for polymers crystallized from the *ordered* melt. Crystallization in the hexagonal-packed cylinder phase has been investigated for a series of PE-poly(3-methyl-1-butene) diblocks [14,16]. Crystallization occurs within the cylinders if the sample is cooled rapidly from the melt, whereas slow cooling results in disruption of the cylindrical mesophase and formation of a lamellar structure. A combination of SAXS and WAXS was used to determine the chain orientation with respect to the cylinders. For low molecular weight samples where chain diffusion is rapid, the chain stems were perpendicular to the cylinder axis, as illustrated in Fig. 11. However, when chain mobility is limited, for example, at higher molecular weight or in diblocks with a glassy PVCH block, chains were tilted with respect to the plane normal to the cylinder axis [16].

7
Block Copolymers with Two Crystallizable Blocks

An intriguing question is what happens upon crystallization in a block copolymer with two crystallizable blocks? This has been investigated for poly(ε-caprolactone)-poly(ethylene oxide)-poly(ε-caprolactone) (PCL-PEO-PCL) triblocks using SAXS, WAXS and DSC [43]. Both components in these copolymers melt around 40–50 °C. The WAXS results revealed that the crystals of the PEO and PCL coexist independently and that there are no eutectic or mixed crystals containing both PEO and PCL. The degree of crystallinity, extracted from the WAXS data, of one of the blocks decreases to zero when it composes less than 25% of the block copolymer. In contrast, in a PCL-PEO binary blend, the two components crystallize independently even with less than 25% of the minor component. Furthermore, in the blend, the lamellar domain spacing and melting temperature did not change with the mixture composition and corresponded to those of the coexisting homopolymers, which was not the case for the block copolymers. These results were interpreted on the basis of a model of independent crystallites of PEO and PCL in the homopolymer blend, segregated due to crys-

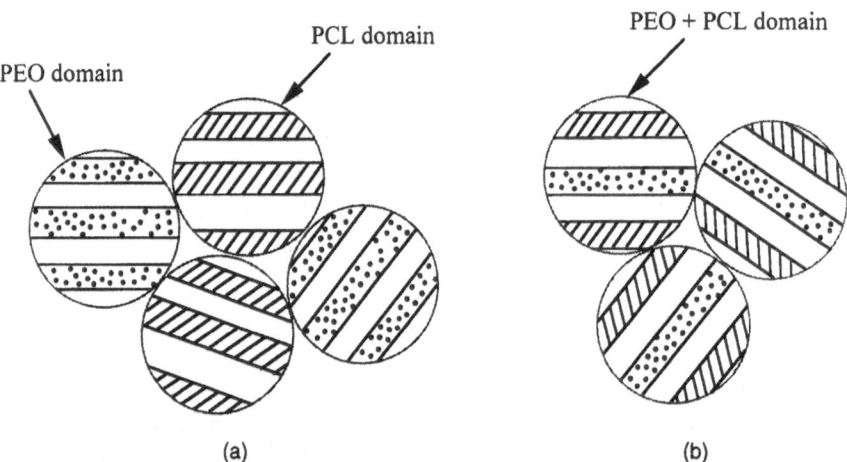

Fig. 12. Schematic illustration of the morphology formed by blends and copolymers of two crystallizable polymers (after [43]). **a** a PEO/PCL blend, **b** in a PCL-PEO-PCL triblock. In the blend, PEO and PCL are phase separated into domains in which each homopolymer crystallizes in a lamellar texture. In the copolymer, PEO and PCL blocks crystallize in the same domain due to chain connectivity

tallization (as observed for other crystalline/crystalline polymer blends) (Fig. 12). In contrast, in the block copolymer, chain connectivity leads to crystalline PEO and PCL lamellae occupying the same crystallite when comparable fractions of each are present (Fig. 12). This leads to the observed large reduction of lamellar thickness and the imperfectness of crystals (as evidenced by the melting temperature depression) and to the WAXS patterns corresponding to coexisting crystalline components [43].

8
Solution Crystallization

Crystallization in or from solution brings another variable, solvent selectivity, into play. If the solvent is selective for the crystalline block, it can swell the crystalline lamellae (T_m is obviously also reduced). In contrast, if the solvent is selective for the non-crystalline block, it can precipitate out of solution in a non-equilibrium structure.

The crystallization of PEO in a PEO-PS-PEO triblock, a PEO-PPO-PEO triblock and a (PPO-PEO)$_4$ four arm starblock in preferential solvents was investigated by Skoulios et al. [44]. In dry copolymers and in a poor solvent for PEO they observed crystallization of the PEO blocks. The solvent was found to be located in the PEO layers in aqueous solution, whereas in selective solvents for PS and PPO it was located in the corresponding block structures [44].

Crystallization from solution of PS-PEO diblocks has been investigated by several groups [45–48]. The principal crystal habits of PS-PEO diblocks and their morphological variations have been described and related to the unit cell of PEO [47]. TEM was also used to probe crystal morphology in these diblocks [48]. Crystallization occurred in square tablets or long strips of layers, the latter breaking up into square-shaped multilayer crystals, often consisting of twisted spiral terraces. PEO homopolymer also forms square tablet single crystals when crystallized from solution (and also hexagonal single crystals). Layer thickness measurements from electron micrographs and SAXS supported a model of chain-folded crystalline PEO lamellae sandwiched between amorphous PS [48].

The degree of crystallinity and chain folding in PS-PEO diblocks has been studied as a function of concentration of diethyl phthalate, which is a selective solvent for PS [45]. Considering copolymers at a fixed concentration, the PEO melting temperature decreases with a reduction in PEO content in the copolymer, as might be expected. At low temperatures a lamellar crystal structure was observed using SAXS whether the high temperature mesophase (i.e. dry or swollen melt) structure was lamellar or hexagonal-packed cylinders. Representative phase diagrams are shown in Fig. 13. PEO was found (via SAXS measurements of the layer thickness) to be highly folded, with up to 28 folds for one diblock. Folding with stems perpendicular to the lamellar interface was assumed. The number of folds, and the interfacial area per PS block, was found to increase with increasing molar mass of the PS block. The effect of concentration of a selective solvent for PS on PEO crystallization has also been explored [45]. For a

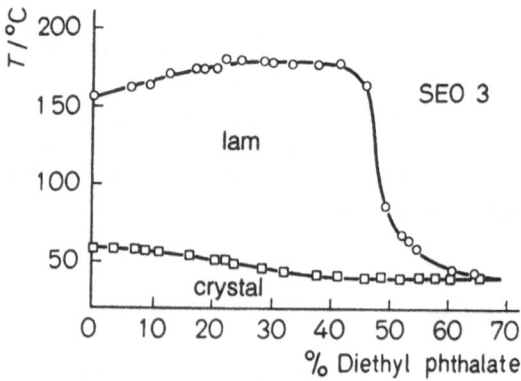

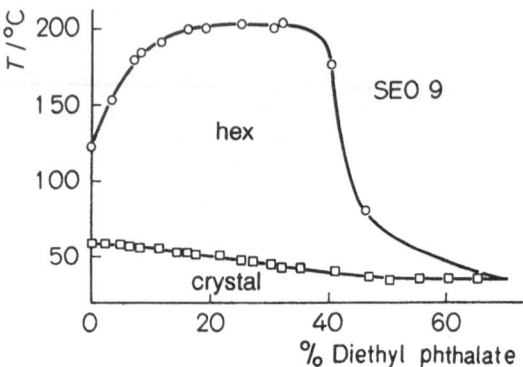

Fig. 13. Phase diagrams for PS-PEO diblocks in diethyl phthalate [45], a selective solvent for PS. SEO-3: M_n=34.7 kg mol^{-1}, 59% PEO. This sample forms a noncrystalline lamellar phase (L) and a lamellar crystalline phase (L_c); SEO-9: M_n=75.0 kg mol^{-1}, 70.5% PEO. This sample forms a noncrystalline hexagonal-packed cylinder phase (H) and an L_c phase

given copolymer at a fixed temperature the number of folds increases (thus the PEO layer thickness decreases) as the solvent concentration increases. At the same time, the PS layer swells and the PS interfacial area increases. Crystallization of PS-PEO diblocks in a selective solvent for the PEO block has been investigated for PS-PEO [46] and PB-PEO [49]. For both systems, a lamellar crystal structure is found below about 45 °C for solvent concentrations ranging from zero to a value characteristic of the copolymer. In these materials, PEO crystallizes in two layers separated by solvent but as the solvent concentration increases, the solvent layer gets thicker, separating the PEO layers but without dissolving them. Above a critical PEO layer thickness (50 Å), increasing solvent concentration leads to discontinuous decrements in the PEO layer thickness due to step increases in the number of chain folds, whilst the degree of crystallinity decreases.

Semicrystalline diblocks in dilute solutions of a solvent selective for the non-crystalline block can form platelet structures. These consist of crystalline chains folded within lamellae between solvated domains of the amorphous block. This constitutes a model system of tethered chains at a flat interface [50]. Self-consistent field theory was used to model the density profile of the tethered chains and SANS and SAXS were performed to provide volume fraction profiles and crystal domain thicknesses, which were compared to the theoretical predictions [50]. The core thickness is due to a balance of an entropic contribution from brush stretching and an enthalpic term from crystalline chain folding (and defects due to ethyl branches). Measurements were performed on solutions of PS-PEO in cyclopentane or PE-PEP in decane. The latter system has also been investigated using SANS by Richter et al. [51]. An additional superstructure was identified, specifically macroaggregates of lamellae, resulting from van der Waals interactions between lamellar sheets [51]. These macroaggregates are needle-like and can be seen by phase contrast optical microscopy. Subsequent work explored the effect of copolymer architecture, via experiments on PE_nPEP_m miktoarm (mixed arm) starblocks [52]. With increasing PEP molecular weight, the extension of the PEP chains in the corona and the reduction in core thickness differed from that expected for a diblock. It was also shown that the platelets can be modeled as disks of diameter ~1 μm (and PE core thickness ~4–10 nm).

9
Crystallization Kinetics

The kinetics of crystallization in polymers are usually analyzed using an Avrami equation [53]. The Avrami theory is based on the nucleation and growth of crystallites, and is not specific to polymers. For athermal crystallization it is assumed that all crystal nuclei are formed and start to grow at time $t=0$; the crystallites then grow at a constant rate until their boundaries meet leading to the formation of spherulites [54]. In thermal crystallization, crystallites are nucleated at a constant rate in space and time. For athermal and thermal crystallization, the initial radial growth of spherulites occurs during primary crystallization. This is followed by the slower process of secondary crystallization, where crystal thickening behind the crystal front occurs, together with the formation of subsidiary crystal lamellae and an increase in crystal perfection [54]. Nucleation and growth via different mechanisms can be described using the Avrami equation:

$$1-\phi=\exp(-kt^n) \tag{2}$$

where ϕ is the relative degree of crystallinity, and k and n are constants which depend on the nucleation and growth mechanism.

The kinetics of crystallization of PE-containing diblocks have been studied using simultaneous SAXS and WAXS [9,11]. The development of PE crystallites (with an enhanced scattering contrast compared with the initial melt state) was followed via the SAXS invariant using Eq. (2). For PE-PEE diblocks (and PE

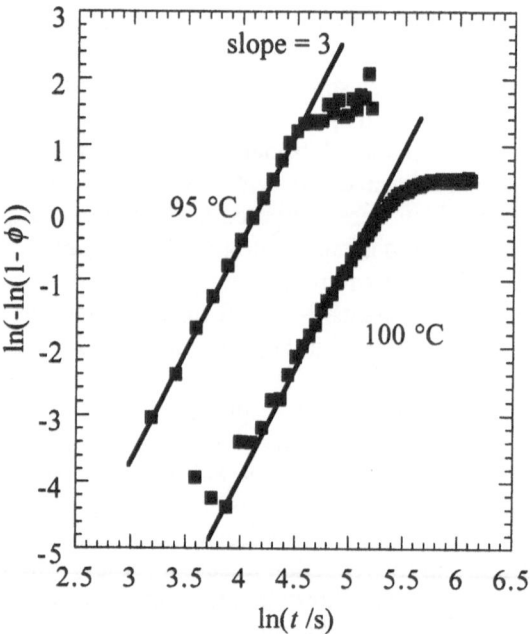

Fig. 14. Avrami plots showing the growth of the relative degree of crystallinity (ϕ) for a PE-PEE diblock (M_n=23 kg mol^{-1}, f_{PE}=0.49) at 95 and 100 °C [11]. The double logarithm of the relative degree of crystallinity determined from the SAXS invariant is plotted against the logarithm of the time

homopolymers) an exponent n=3 has been reported [9,11] (see Fig. 14). This is consistent with the athermal growth of spherulites in three dimensions, as usually observed for the growth of lamellae in crystallizing polymers. However, it was found that this exponent does not describe the secondary crystallization process. A PE-PVCH diblock differed from the other samples due to the glassy component and the kinetics of crystallization were found to be much slower. It was surmised that the presence of a glassy component inhibits nucleation and growth of spherulites (and leads to a suppression of the crystallization temperature) and instead leads to heterogeneously nucleated two-dimensional growth which is also characterized by n=3 [9].

The dynamics of crystallization from the homogeneous melt in strongly segregated PE-PEP and PE-PEE diblocks have been studied using SAXS, WAXS and DSC [55]. Up to four orders of reflections in the SAXS profiles developed rapidly and simultaneously during isothermal crystallization. This demonstrates that a domain structure with sharp interfaces develops even for small undercoolings. During the rapid primary stage of crystallization, the SAXS peak positions are constant, suggesting that the crystalline structures nucleate and grow to fill the sample without any internal rearrangement. This contrasts with PE homopolymers which show substantial changes in the crystallite size distribution in this stage. The simultaneous evolution of SAXS and WAXS data for the diblocks shows that microstructure and crystallinity develop with identical growth kinetics. At longer times, following primary crystallization, the SAXS and WAXS intensities continued to grow in parallel but at a much reduced rate. The parallel growth of SAXS and WAXS intensities indicates that the development of regions of crystalline material occurs in tandem with the creation of new lamellar stacks [55]. Similar results were reported for crystallization from the ordered melt of a PE/head-to-head poly(propylene) diblock [56]. The peak shape and position during crystallization could be adequately approximated by a linear combination of the peaks observed before and after crystallization, consistent with a nucleation and growth process where melt is simply converted to crystallized material with no subsequent change in morphology. Nojima et al. investigated crystallization in PCL-PB diblocks via SAXS [12,13,57]. They also concluded that for samples crystallizing from the ordered melt, the increase in the crystalline morphology was accompanied by a simultaneous decrease in the melt structure [12]. Here, crystallization of the PCL block drives the primary process, irrespective of the microphase separated structure. In the late stage, the crystallization of the diblocks was found to be significantly retarded with respect to the homopolymer, with a more gradual increase in crystallinity compared to the latter [13,57], in accord with work on PE-containing diblocks [9,11].

Crystallization-induced phase separation can occur for concentrated solutions (gels) of diblocks [58,59]. SAXS/WAXS experiments on short PM-PEO [PM=poly(methylene) i.e. alkyl chain] diblocks revealed that crystallization of PEO occurs at low temperature in sufficiently concentrated gels (>ca. 50% copolymer). This led to a semicrystalline lamellar structure coexisting with the cubic micellar phase which can be supercooled from high temperatures where PEO is molten. These experiments on oligomeric amphiphilic diblocks establish a connection to the crystallization behaviour of related nonionic surfactants.

10
Theories for Crystallization

Ashman and Booth [26] described the application of the Flory–Vrij theory [30] to semicrystalline block copolymers. The Flory–Vrij theory allows for partial melting of the crystalline lamellae with the formation of an amorphous layer containing chain ends (i.e. less than perfect crystallinity). It is based on a probabilistic model for chain folding. For block copolymers, the end interfacial energy was found to be $\sigma_e \approx \sigma_a + \sigma_0$, where σ_a is the free energy of formation of the amorphous layer from the melt ($\sigma_a \sim 3.5$ kJ mol^{-1}) and σ_0 is the free energy of formation of the crystalline/amorphous interface ($\sigma_0 \sim 3$ kJ mol^{-1}) [26]. This model can also be used to obtain estimates for the degree of crystallinity, as well as end interfacial energies.

DiMarzio et al. developed a scaling model for semicrystalline lamellar structures with alternating layers of amorphous and crystalline blocks, with crystalline chains folded perpendicular to the interface [24]. The model system includes solvent in the amorphous phase. The theory makes predictions for the scaling of amorphous and crystalline domain thicknesses with degree of polymerization. These, however, have yet to be tested for crystallization in solution (comparisons have been made for experiments on solvent-free copolymers, see Sect. 4). It was emphasized that the tendency for chains to be extended in the crystalline region is opposed by the preference for coiled chains in the amorphous block, and that these opposing tendencies lead to an equilibrium degree of chain folding in the crystalline layer [24]. The model allows for the stretching of polymer chains, the change in packing entropy arising from changes in orientation of bonds, and the space-filling properties of the chains. It is purely an entropic model, with no allowance for the enthalpic mixing terms considered in the Flory–Vrij theory.

A more comprehensive theory for the thermodynamics of semicrystalline diblocks has been developed using self-consistent mean field theory applied to diblocks with one amorphous block and one crystallizable block [20].The amorphous regions were modelled as flexible chains, and the crystalline regions as folded chains. Both monolayers and bilayers of once-folded chains were considered. Expressions were derived for the thickness of the amorphous and crystalline region and the number of folds. The central result is the domain spacing scaling [Eq. (1)].

Acknowledgements. Special thanks are due to Professor Tony Ryan and Dr Patrick Fairclough (University of Sheffield) with whom I have collaborated on much of the work discussed in this review. I am also grateful to the Engineering and Physical Sciences Research Council (UK) for funding.

References

1. Hamley IW (1998) The physics of block copolymers. Oxford University Press, Oxford
2. Mohajer Y, Wilkes GL, Wang IC, McGrath JE (1982) Polymer 23:1523
3. Hirata E, Ijitsu T, Soen T, Hashimoto T, Kawai H (1975) Polymer 16:249
4. Séguéla R, Prud'homme J (1989) Polymer 30:1446
5. Cohen RE, Cheng P-L, Douzinas K, Kofinas P, Berney CV (1990) Macromolecules 23:324
6. Veith CA, Cohen RE, Argon AS (1991) Polymer 32:1545
7. Khandpur AK, Macosko CW, Bates FS (1995) J Polym Sci Polym Phys Ed 33:247
8. Vonk CG (1973) J Appl Cryst 6:81
9. Hamley IW, Fairclough JPA, Bates FS, Ryan AJ (1998) Polymer 39:1429
10. Balta-Calleja FJ, Vonk CG (1989) X-ray scattering of synthetic polymers. Elsevier, London
11. Ryan AJ, Hamley IW, Bras W, Bates FS (1995) Macromolecules 28:3860
12. Nojima S, Kato K, Yamamoto S, Ashida T (1992) Macromolecules 25:2237
13. Nojima S, Nakano H, Takahashi Y, Ashida T (1994) Polymer 35:3479
14. Quiram DJ, Register RA, Marchand GR (1997) Macromolecules 30:4551
15. Quiram DJ, Register RA, Marchand GR, Ryan AJ (1997) Macromolecules 30:8338
16. Quiram DJ, Register RA, Marchand GR, Adamson DH (1998) Macromolecules 31:4891
17. Hamley IW, Fairclough JPA, Terrill NJ, Ryan AJ, Lipic PM, Bates FS, Towns-Andrews E (1996) Macromolecules 29:8835
18. Hamley IW, Fairclough JPA, Ryan AJ, Bates FS, Towns-Andrews E (1996) Polymer 37:4425

19. Ryan AJ, Fairclough JPA, Hamley IW, Mai S-M, Booth C (1997) Macromolecules 30:1723
20. Whitmore MD, Noolandi J (1988) Macromolecules 21:1482
21. Douzinas KC, Cohen RE, Halasa AF (1991) Macromolecules 24:4457
22. Rangarajan P, Register RA, Fetters LJ (1993) Macromolecules 26:4640
23. Unger R, Beyer D, Donth E (1991) Polymer 32:3305
24. DiMarzio EA, Guttman CM, Hoffman JD (1980) Macromolecules 13:1194
25. Nojima S, Yamamoto S, Ashida T (1995) Polym J 27:673
26. Ashman PC, Booth C (1975) Polymer 16:889
27. Ashman PC, Booth C, Cooper DR, Price C (1975) Polymer 16:897
28. Booth C, Pickles CJ (1973) J Polym Sci Polym Phys Ed 11:259
29. Flory PJ (1949) J Chem Phys 17:223
30. Flory PJ, Vrij S (1963) J Am Chem Soc 85:3548
31. Viras F, Luo Y-Z, Viras K, Mobbs RH, King TA, Booth C (1988) Makromol Chem 189:459
32. Al Kafaji JKH, Booth C (1981) Makromol Chem 182:3671
33. Booth C, Domszy RC, Leung Y-K (1979) Makromol Chem 180:2765
34. Cooper DR, Leung Y-K, Heatley F, Booth C (1978) Polymer 19:309
35. Domszy RC, Mobbs RH, Leung Y-K, Heatley F, Booth C (1979) Polymer 20:1204
36. Domszy RC, Mobbs RH, Leung Y-K, Heatley F, Booth C (1979) Polymer 21:588
37. Yang Y-W, Tanodekaew S, Mai S-M, Booth C, Ryan AJ, Bras W, Viras K (1995) Macromolecules 28:6029
38. Mai S-M, Fairclough JPA, Viras K, Gorry PA, Hamley IW, Ryan AJ, Booth C (1997) Macromolecules 30:8392
39. Honeker CC, Thomas EL (1996) Chem Mater 8:1702
40. Douzinas KC, Cohen RE (1992) Macromolecules 25:5030
41. Cohen RE, Bellare A, Drzewinski MA (1994) Macromolecules 27:2321
42. Kofinas P, Cohen RE (1994) Macromolecules 27:3002
43. Nojima S, Ono M, Ashida T (1992) Polym J 24:1271
44. Skoulios AE, Tsouladze G, Franta E (1963) J Polym Sci C 4:507
45. Gervais M, Gallot B (1973) Makromol Chem 171:157
46. Gervais M, Gallot B (1973) Makromol Chem 174:193
47. Lotz B, Kovacs B (1966) Koll-Z u Z Polym 209:97
48. Lotz B, Kovacs AJ, Bassett GA, Keller A (1966) Koll-Z u Z Polym 209:115
49. Gervais M, Gallot B (1977) Makromol Chem 178:1577
50. Lin EK, Gast AP (1996) Macromolecules 29:4432
51. Richter D, Schneiders D, Monkenbusch M, Willner L, Fetters LJ, Huang JS, Lin M, Mortensen K, Farago B (1997) Macromolecules 30:1053
52. Ramzi A, Prager M, Richter D, Efstratiadis V, Hadjichristidis N, Young RN, Allgaier JB (1997) Macromolecules 30:7171
53. Avrami M (1939) J Chem Phys 7:1103
54. Gedde UW (1995) Polymer physics. Chapman and Hall, London
55. Rangarajan P, Register RA, Adamson DH, Fetters LJ, Bras W, Naylor S, Ryan AJ (1995) Macromolecules 28:1422
56. Rangarajan P, Register RA, Fetters LJ, Bras W, Naylor S, Ryan AJ (1995) Macromolecules 28:4932
57. Nojima S, Nakano H, Ashida T (1993) Polymer 34:4168
58. Hamley IW, Pople JA, Ameri M, Attwood D, Booth C, Ryan AJ (1998) Macromol Chem Phys 199:1753
59. Hamley IW, Pople JA, Ameri M, Attwood D, Booth C (1998) Coll Surf A 145:185
60. Lovinger AJ, Han BJ, Padden FJ, Mirau PA (1993) J Polym Sci Polym Phys Ed 31:115

Editor: Prof. R.J. Young
Received: August 1998

Viscoelastic Behavior of Epoxy Resins Before Crosslinking

Tsuneo Koike

R&D Laboratory, Yuka Shell Epoxy Co., Ltd., No.1 Shiohama-cho, Yokkaichi,
Mie 510–0851 Japan
E-mail: koiket@anet.ne.jp

A review of current viscoelastic studies is presented for epoxy resins before crosslinking. Epoxy resins are usually crosslinked with adequate curing agents and applied to various commercial uses as thermoset resins after crosslinking. Some epoxy resins, however, have been used without curing agents and treated as thermoplastic resins. The oligomers of the diglycidylether of bisphenol-A (DGEBA) are the most popular epoxy resins. This article summarizes and discusses viscoelastic properties, such as the melt viscosity, the dielectric relaxation time, and the ionic conductivity, for the uncured DGEBA oligomer on the basis of free-volume concept.

Advances in Polymer Science, Vol.148
© Springer-Verlag Berlin Heidelberg 1999

Abbreviations

α	distribution parameter of Havriliak-Negami equation
α_f	thermal expansion coefficient of the free volume
α_g	thermal expansion coefficient below T_g
α_l	thermal expansion coefficient above T_g
β	skewness parameter of Havriliak-Negami equation
β'	parameter of KWW relaxation function
C_1, C_2	WLF parameters for the melt viscosity (η)
C_1', C_2'	WLF parameters for the dielectric relaxation time (τ)
C_1'', C_2''	WLF parameters for the direct current conductivity (σ)
D	constant in Vogel-Fulcher equation
DC	direct current
DDM	4,4'-diaminodiphenylmethane
DDS	4,4'-diaminodiphenylsulfone
DGEBA	diglycidyl ether of bisphenol-A
DSC	differential scanning calorimetry
ε^*	complex dielectric constant
ε'	dielectric constant
ε''	dielectric loss
ε_0	low-frequency limit to ε'
ε_∞	high-frequency limit to ε'
ε_d''	dipolar component of ε''
ε_i''	ionic component of ε''

e	base of natural logarithms
E_0	permittivity of free space (8.855×10^{-14} F/cm)
ECH	epichlorohydrin
f_g	free-volume fraction at T_g
GPC	gel permeation chromatography
ϕ	KWW relaxation function
γ	numerical factor introduced to correct for overlap of free-volume
η	melt viscosity
HPLC	high performance liquid chromatography
i	square root of -1
k	exponent in $\sigma\eta^k$=const
k_B	Boltzmann constant
KWW	Kohlrausch-Williams-Watts
l	exponent in η/τ^l=const
m	exponent in $\sigma\tau^m$=const
μ	mobility of ion
$\bar{M}_n$	number-average molecular weight
$\bar{M}_w$	weight-average molecular weight
n	repeated number in structural formula
N	number of ion
NMR	nuclear magnetic resonance
PACM	bis(p-aminocyclohexyl) methane
PAS	positron annihilation spectroscopy
PGEBAN	poly(glycidyl ether) of bisphenol-A novolac
PVAc	poly(vinyl acetate)
Q	charge of ion
R	radius of ion
σ	direct current (ionic) conductivity
τ	dielectric relaxation time
τ_0	constant in Vogel-Fulcher equation
t	time
T	temperature
T_g	glass transition temperature
T_0	constant in Vogel-Fulcher equation
TBA	torsional braid analysis
TGDDM	tetraglycidyl 4,4'-diaminodiphenylmethane
TTT	time-temperature-transformation
υ^*	critical volume for the transport of a moving unit
υ^*_η	υ^* for the chain segment
υ^*_σ	υ^* for the ionic charge carrier
υ^*_τ	υ^* for the dipole
$\bar{\upsilon}_m$	average molecular volume
ω	angular frequency
WLF	Williams-Landel-Ferry
ζ	friction constant

1
Introduction

Epoxy resins are regarded as thermosetting resins and have found various commercial applications after crosslinking with adequate curing agents [1–3]. However, some epoxy resins have been used as thermoplastic resins without curing agents. Figure 1 shows the applications of epoxy resins that are classified to three categories: thermosets in combination with curing agents, thermoplastics without curing agents, and raw materials for modification. The use in thermoplastics is not popular compared with the two other applications. Typical thermoplastic applications are found in stabilizers for vinyl resins, toners for copying machines, fire retardants for engineering plastics, and sizing material for glass or carbon fibers.

The epoxy resin most frequently used is the oligomer of the diglycidyl ether of bisphenol-A (DGEBA) whose chemical structure is shown below [1–3].

$$CH_2\text{-}CH\text{-}CH_2\text{-}O\left[\text{-}\bigcirc\text{-}\overset{\overset{\displaystyle CH_3}{|}}{\underset{\underset{\displaystyle CH_3}{|}}{C}}\text{-}\bigcirc\text{-}O\text{-}CH_2\text{-}\underset{\underset{\displaystyle OH}{|}}{CH}\text{-}CH_2\text{-}O\right]_n\text{-}\bigcirc\text{-}\overset{\overset{\displaystyle CH_3}{|}}{\underset{\underset{\displaystyle CH_3}{|}}{C}}\text{-}\bigcirc\text{-}O\text{-}CH_2\text{-}CH\text{-}CH_2$$

The DGEBA is composed of linear molecules with different molecular weights according to the variation of the repeated number (n) in the structural formula.

A large number of evaluation studies have been carried out for the DGEBA oligomers that were cured with various curing agents. These studies are performed to obtain practical data needed in the actual application of the crosslinked DGEBA oligomer. In contrast, only a small number of investigations are focused on the viscoelastic study of the uncured DGEBA oligomer, so that the nature of the DGEBA oligomer is still not well understood. This article reviews current viscoelastic studies on uncured DGEBA oligomers for a basic understanding of the uncured oligomer. The DGEBA oligomer has a molecular weight between that of a monomer and a polymer. The viscoelastic study of the oligomer may also provide valuable information for a better understanding of the nature of the polymer or monomer.

The crosslinked epoxy resins, as well as the epoxy resin system during curing, have been analyzed by many researchers. The epoxy resin curing mechanism, however, is not always fully defined because of a variety of interactions among the components in the curing system. The uncured linear DGEBA oligomer is considered to be a simple structural model for the reactive DGEBA-curing agent system before gelation during curing. Viscoelastic studies of DGEBA oligomers are also summarized in this article because of their usefulness in the analysis of the epoxy resin curing system.

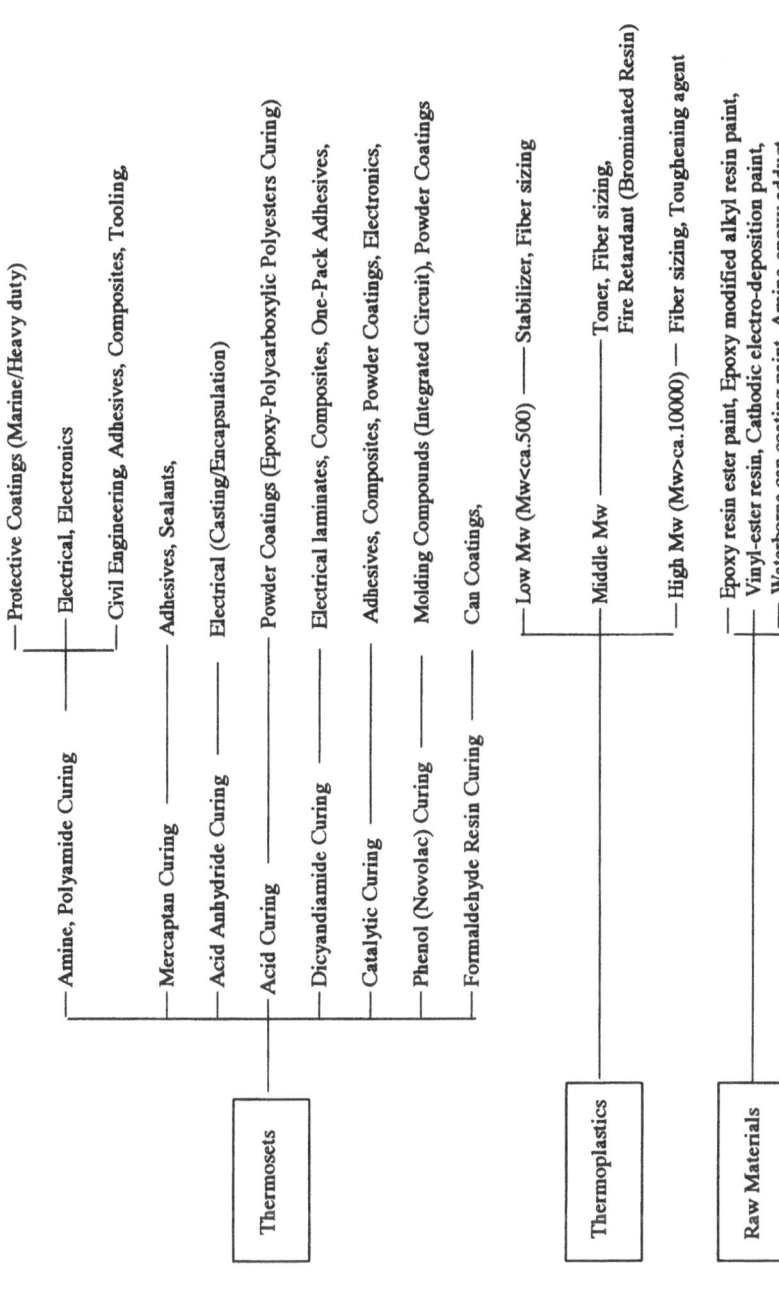

Fig. 1. Applications of epoxy resins

2
Bisphenol-A Type Epoxide Oligomers

An epoxy resin is a polymeric material containing, on the average, two or more epoxide groups in the molecule. The most widely used epoxy resin is the diglycidyl ether of bisphenol–A (DGEBA). The DGEBA can be prepared mainly by two routes, the conventional and advancement processes [1–3]. The conventional-type resin is manufactured by a direct addition of epichlorohydrin (ECH) to bisphenol–A in the presence of sodium hydroxide. Figure 2 shows the process flow of a typical conventional production process. The molecular weight of the product is governed by the ratio of ECH to bisphenol-A. A lower molecular weight DGEBA oligomer is produced with a larger molar ratio of ECH to bisphenol-A. The conventional process is also called the taffy or the solvent–finished process. Another production route, the advancement or fusion process, involves a reaction of low molecular weight DGEBA with bisphenol-A in order to advance the molecular weight as shown below [1–3].

low molecular weight DGEBA   bisphenol-A

+ catalyst

advancement-type DGEBA

The molecular weight of the advancement-type resin is controlled by the ratio of bisphenol-A to the low molecular weight DGEBA. A higher molecular weight oligomer can be produced with a higher molar ratio of bisphenol-A to a low molecular weight DGEBA. The repeating number n of commercially available DGEBA varies from near zero up to around 18. DGEBA is called an epoxide oligomer. Branching in the DGEBA oligomer occurs during the production processes previously mentioned. The degree of side chain branching has been studied using nuclear magnetic resonance (NMR), gel permeation chromatography (GPC), or high performance liquid chromatography (HPLC). The extent of branching is very small as reported by Batzer and Zahir [4], Bantle and Burchard [5], and Eppert et al. [6].

A. Reaction mechanism

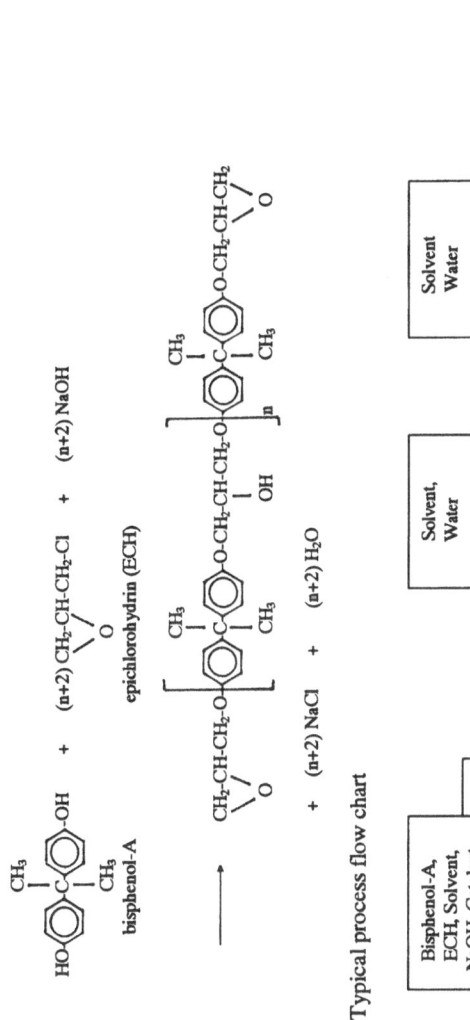

B. Typical process flow chart

Fig. 2. A conventional production process of bisphenol-A type epoxide oligomer

Table 1. Properties of DGEBA oligomers [10]

DGEBA Oligomer	T_g (°C)	$\bar{M}_n$	$\bar{M}_w$	$\bar{M}_w/\bar{M}_n$	n^a	Na+[b] (ppm)	Cl−[c] (ppm)
Epikote 828	−16	366	388	1.06	0.1	<1	<1
Epikote 834	0	479	590	1.23	0.5	1	<1
Epikote 1001	30	898	1396	1.55	2.0	<1	<1
Epikote 1002	39	1147	1891	1.63	2.8	1	<1
Epikote 1004	54	1538	2640	1.72	4.2	2	<1

[a] Repeated number in the structural formula
[b] Measured by atomic absorption analysis
[c] Measured by potentiometric titration with 0.01 N silver nitrate

The DGEBA oligomer generally exhibits a molecular weight distribution. The ratio of the weight-average molecular weight ($\bar{M}_w$) to the number-average molecular weight ($\bar{M}_n$), $\bar{M}_w/\bar{M}_n$, is an indication of the molecular weight distribution. The $\bar{M}_w$ and $\bar{M}_n$ values are usually determined by GPC using calibration curves based on polystyrene standards. Mori [7] reported a practical calibration method for three DGEBA oligomers ($2900 \leq \bar{M}_w \leq 15300$). Characterizations including $\bar{M}_w/\bar{M}_n$ values for conventional-type DGEBA oligomers are listed in Table 1.

Both conventional- and advancement-type DGEBA oligomers are described by the same structural formula. Some differences, however, are observed between the two types of DGEBA oligomers [3, 8, 9]. A major difference is the pattern of the molecular weight distribution. Typical GPC charts are given in Fig. 3 for the conventional- and advancement-type DGEBA oligomers, Epikote 1002 ($\bar{M}_w$=1891) and Epikote 1002F ($\bar{M}_w$=2111), respectively. The conventional-type resin has both even- and odd-numbered oligomers, whereas in the advancement-type resin the even-numbered oligomers (n=0, 2, 4, 6, etc.) are dominant [3]. Another difference is found in the concentration of the ionic component. The advancement-type oligomer contains a residue of the catalyst that is not removed during the production process. Low molecular weight ionic products, such as basic inorganic compounds, amines, and quaternary ammonium salts are typical catalysts [3]. The catalyst is usually added in a concentration of tens to hundreds of ppm in a commercially available advancement-type DGBEA oligomer, which has a significant influence on the ionic conduction measurement. On the other hand, the DGEBA oligomer based on the conventional process experiences little effect by the catalyst because the catalyst is removed in the washing process shown in Fig. 2. The major ions in the conventional-type DGEBA oligomers are Na$^+$ and Cl$^-$ due to the raw materials used [1–3]. The concentrations of these two ions in the conventional-type oligomer are very low (in ppm order of magnitude or less) as shown in Table 1 [10]. Other differences are observed in the concentrations of minor terminal groups, such as saponifiable chlorine, α-glycol, and phenolic hydroxyl groups [3]. The concentrations of these terminal groups are very low, which has little influence on the viscoelastic properties.

A. Conventional Type
   Epikote 1002 (Mw=1891)

B. Advancement Type
   Epikote 1002F (Mw=2111)

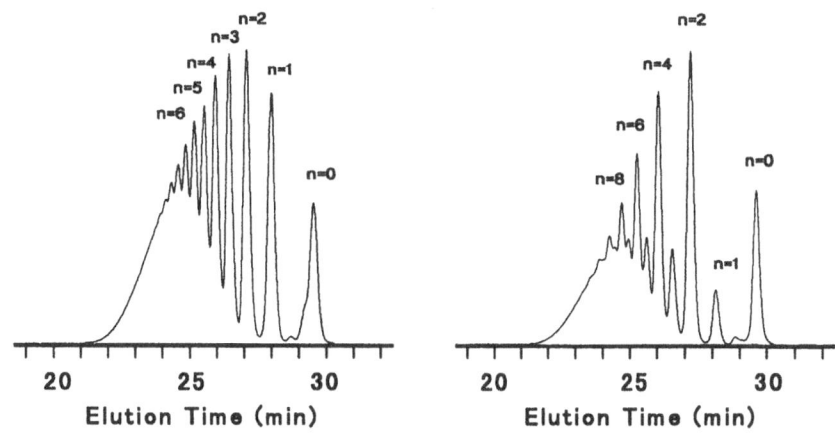

Fig. 3. Gel permeation chromatograms of two DGEBA oligomers

A polymer having the bisphenol-A backbone structure is called a phenoxy resin. Its chemical structure is as follows:

$$\left[\!-O\!-\!\bigcirc\!-\!\underset{\underset{CH_3}{|}}{\overset{\overset{CH_3}{|}}{C}}\!-\!\bigcirc\!-\!O\text{-}CH_2\text{-}\underset{\underset{OH}{|}}{CH}\text{-}CH_2\!-\!\right]_n$$

The commercially available phenoxy resin has $\bar{M}_w$ around 50,000 and little epoxy functionality. The phenoxy polymer is then classified as a polyol or a poly(hydroxy ether) [3]. Typical applications for the phenoxy resin are as hot-melt adhesives, coatings for beverage cans, and toughening agents for various epoxy resin formulations. The viscoelastic properties of phenoxy resins have been studied by Alegría et al. [11–15]. The phenoxy polymer can be used as a reference in the viscoelastic study of DGEBA oligomer.

3
Characterization Methods for Epoxide Oligomers

Epoxide oligomers before crosslinking are characterized by a number of properties, such as epoxy content, melt (or solution) viscosity, volatile content, melting (or softening) point, color, or density [1–3]. These characterizations are usually made for quality control or the specification of products. In addition, GPC and HPLC are frequently used for studying the molecular composition or purity

Table 2. Typical evaluation methods for epoxy resin curing systems

Properties	Measuring technique	Measurable stage[a]	Ref.
1. Functional group	Fourier Transfer Infrared Spectroscopy (FT-IR), Nuclear Magnetic Resonance (NMR)	A→C	17–20
2. Gel fraction	Gel Permeation Chromatography (GPC), Extraction	A→C	21
3. Molecular weight	GPC, High Performance Liquid Chromatography (HPLC)	A→B	22–23
4. Reaction heat	Differential Scanning Calorimetry (DSC)	A→C	24–25
5. Viscosity	Viscometer, Dynamic Mechanical Analysis (DMA)	A	26–29
6. Flow properties	Parallel plate plastometer, Disc-cure method	A→B	30, 31
7. Torque	Curelastometer, Cycloviscograph	A→B	32, 33
8. Hardness	Hot Rockwell hardness, Pencil hardness	B→C	34
9. Needle penetration	Höppler consistometer, Needle track tester	A→C	34
10. Specific volume	Cure shrinkage measurement (Dilatometry, etc.)	A→C	35, 36
11. Viscoelasticity	Torsional Braid Analysis (TBA), DMA, Dynamic Spring Analysis (DSA)	A→C	37–39
12. Glass transition temperature	DSC, DMA	A→C	40, 41
13. Mechanical strength	Adhesion strength, Flexural strength	B→C	21
14. Dielectric properties	Dielectric measurement, Microelectrode method	A→C	42–45
15. Conductivity	Direct current measurement, Dielectric measurement, Thermally Stimulated Discharge (TSD) method	A→C	46–49
16. Ultrasonic properties	Ultrasonic resonance method, Fokker Bond Tester	A→C	50, 51
17. Free volume	Positron Annihilation Spectroscopy (PAS)	A→C	52
18. Photoluminescence	Fluorescence monitoring	A→C	53, 54
19. Appearance	Electron microscope, Video camera observation	A→C	55, 56

[a] Measurable stage: "A" stage (uncured) → "B" stage (before gelation) → "C" stage (cured)

of epoxide oligomers [4–9]. Infrared spectroscopy and NMR are utilized to determine the chemical structure of epoxide oligomers as described in a review by Mertzel and Koenig [16].

Uncured epoxide oligomers are also investigated to obtain basic information for an analysis of reactive epoxy resin systems during curing. Typical evaluation methods [17–56] for curing systems are summarized in Table 2. Among the evaluation methods applicable to the cure analysis, the methods for measuring the following viscoelastic properties are considered important from the standpoint of the on-line monitoring of the curing system:

a) melt viscosity
b) dielectric relaxation time
c) direct current (DC) or ionic conductivity

These viscoelastic properties above the glass transition can be correlated with each other as the transport phenomenon in the same oligomer. The relationships between two of the these three properties previously listed are also important for a basic understanding of the nature of the uncured epoxide oligomer.

4
Viscoelastic Properties of Epoxide Oligomers

In this section, the melt viscosity, the dielectric relaxation time, and the DC conductivity are reviewed and summarized for DGEBA oligomers before crosslinking in terms of their temperature-dependent behavior.

4.1
Melt Viscosity

Flow properties have been studied for DGEBA oligomers before crosslinking by several researchers [57–62]. Aleman [57,58] discussed shear-rate effects on the shear viscosity measured for some DGEBA oligomers and found that each oligomer exhibited Newtonian behavior up to the shear rate of 2000 s^{-1}. Ghijsels et al.[59] reported that the temperature dependence of the zero-shear viscosity was well described by the Vogel equation. Utracki and Gihjsels [60] analyzed the melt viscosity of several DGEBA oligomers according to the equation based on Doolittle's formula. Table 3 lists some empirical temperature dependences of the melt viscosity for DGBEA oligomers [59–62]. The temperature dependence of the melt viscosity for each DGEBA oligomer is expressed by the Doolittle [63], Vogel [64, 65], Williams-Landel-Ferry (WLF) [66], or Berry & Fox type [67] equations. These four equations are converted into each other on the basis of the free-volume concept.

The melt viscosity of the DGEBA oligomer is measured in many cases by the cone-plate type viscometer [59–62]. Figure 4 shows viscosity-angular frequency curves [61] for Epikote 1002 ($\bar{M}_w$=1891, T_g=39 °C) at 90 and 120 °C. The epoxide oligomer exhibits Newtonian flow behavior up to a relatively high angular

Table 3. Temperature dependence of melt viscosity for DGEBA oligomers

Relationship	Remarks	Ref.
A. Vogel type		
$$\log \frac{\eta(T)}{\eta(1.2T_g)} = \log 38 \times 10^{-6} + \frac{1.38}{(T/T_g - 0.89)}$$	η : melt viscosity T : temperature Tg : glass transition temperature	59
B. Doolittle type		
$\ln \eta = B_0 + B_1 Y$ $Y = 1/\{0.907\exp[0.11515(T/T_g)^{3/2}] - 1\}$	η : melt viscosity B_0, B_1 : constants T : temperature T_g : glass transition temperature	60
C. WLF type		
$$\log \frac{\eta(T)}{\eta(T_g)} = \frac{-C_1(T - T_g)}{C_2 + T - T_g}$$	η : melt viscosity C_1, C_2 : WLF parameters T : temperature T_g : glass transition temperature	61
D. Berry & Fox type		
$$\eta = F_\eta \zeta_0 \exp\left[\frac{B}{f_g + \alpha_1(T - T_g)}\right]$$ $\zeta_0 = 2.49 \times 10^{-5}$ poise-mole/g $f_g/B = 0.025$	η : melt viscosity F_η : structure sensitive factor ζ_0 : inherent fraction factor T_g : glass transition temperature f_g : free volume fraction at T_g B : constant, T: temperature α_1 : thermal expansion coefficient above T_g	62

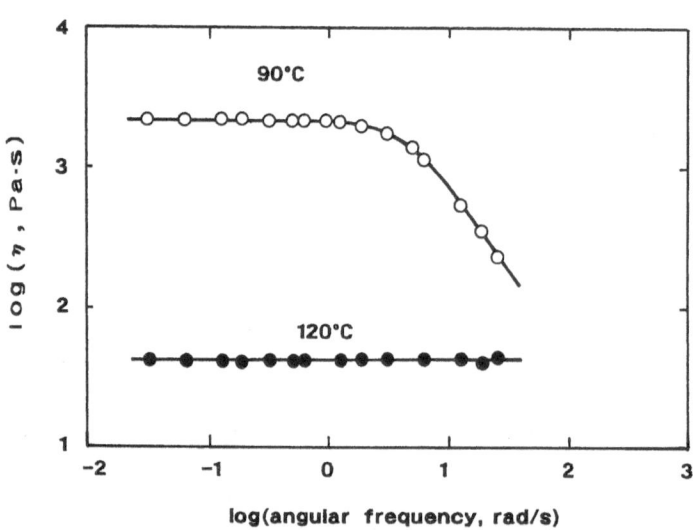

Fig. 4. Angular frequency dependence of melt viscosity for Epikote 1002 at 90 and 120 °C [61]

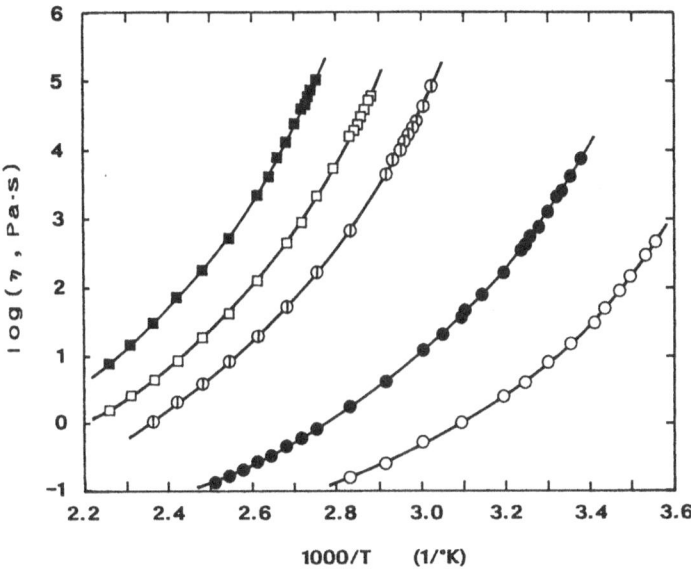

Fig. 5. Temperature dependence of melt viscosity for five DGEBA oligomers: ○ Epikote 828; ● Epikote 834; ① Epikote 1001; ❑ Epikote 1002; ■ Epikote 1004 [61]

frequency (high shear stress) compared with higher molecular weight polymers. The melt viscosity of the oligomer is determined by the value in the Newtonian region. Figure 5 shows the melt viscosity of five DGEBA oligomers ($388 \leq \bar{M}_w \leq 2640$) as a function of the reciprocal Kelvin temperature [61]. The temperature dependencies of the melt viscosity for the five oligomers can be described in the following WLF equation [66]:

$$\log \frac{\eta(T)}{\eta(Tg)} = \frac{-C_1(T - T_g)}{C_2 + T - T_g} \tag{1}$$

where η is the melt viscosity, T is the temperature ($T_g \leq T \leq T_g + 100\ °C$), T_g is the glass transition temperature, and C_1 and C_2 are WLF parameters. The best-fit WLF parameters [61] for the five DGEBA oligomers are listed in Table 4.

Other than the equations listed in Table 3, the following two descriptions have been proposed for the temperature dependencies of the melt viscosity of low $\bar{M}_w$ DGEBA oligomers ($\bar{M}_w \approx 380$) by Marwedel [68], and Hadjistamov [69].

$$\log[\eta(T)] = 6.5083 - 1.4934T^{1/3} \tag{2}$$

$$\eta(T) = 1.8527 \times 10^{10}T^{-4.4136} \tag{3}$$

These approximated functions are considered applicable only to low $\bar{M}_w$ DGEBA oligomers. An Arrhenius-type temperature dependence has been also

Table 4. WLF parameters of DGEBA oligomers [10, 61]

DGEBA Oligomer	T_g (°C)	$\bar{M}_w$	Viscosity		Relaxation time		Conductivity	
			C_1	C_2	C_1'	C_2'	C_1''	C_2''
Epikote 828	−16	388	11.27	25.8	15.69	28.0	12.92	28.0
Epikote 834	0	590	12.23	36.0	16.74	35.5	12.29	33.9
Epikote 1001	30	1396	13.89	45.7	16.14	48.9	10.99	48.3
Epikote 1002	39	1891	15.36	42.2	16.54	48.8	11.05	47.1
Epikote 1004	54	2640	14.99	50.0	15.46	52.6	9.76	54.6

reported for the melt viscosity of a low $\bar{M}_w$ DGEBA oligomer in a relatively narrow range of temperature by Lin and Pearce [70].

4.2
Dielectric Properties

4.2.1
Dielectric Measuring Cell

An epoxide oligomer changes from an amorphous solid to a viscous liquid state above the T_g. The dielectric measurement of the oligomer above the T_g is not easy to perform using the conventional parallel-plate type specimen holder. Several researchers have investigated epoxy resin systems using hand-made parallel-plate cells or the commercially available micro-dielectric sensor (a microchip-type cell). Table 5 shows seven dielectric cells that have been applied to dielectric measurements of epoxy resin systems before or during curing [10, 71–79]. These measuring cells are usually disposable. An example of the parallel-plate cell, a vertical type [10], is shown in Fig. 6. This type of cell makes it possible to measure the sample not only in the liquid but also in the solid state because of the vertical design and simple shape of the cell. A three-terminal electrode in the parallel–plate type cell is preferable to obtain accurate measuring data. It is necessary to take proper measures for preventing the corrosion of the cell by ionic impurities in the epoxide oligomer. The correction for the thermal expansion of the cell is also needed when the dielectric measurement is taken over a wide temperature range above the T_g. The other type of measuring device, a microchip-type cell [78, 79], is the sensor that combines a comb electrode with a pair of field-effect transistors in a silicone-integrated microcircuit. Senturia and Sheppard [79] precisely describe the mechanism of the microdielectrometry that has been widely used for the analysis of the epoxy resin system during curing. The microdielectrometry can be applied to the simultaneous measurement method, for example, the combination of the dielectric and dynamic mechanical analysis, as reported by Gotro and Yandrasits [80], and Simpson and Bidstrup [81]. Other than the microchip cell listed in Table 5, Kranbuehl et al. [82] developed a micro-sensor (1 inch by 1/2 inch in area) called the frequency

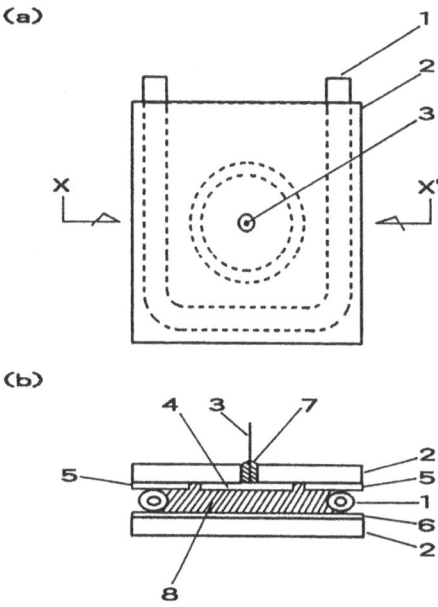

Fig. 6. Configuration of a vertical parallel-plate cell [10]. **a** Side view; **b** Cross section view (X–X'): (1) silicone tube (outer diameter, 4 mm), (2) glass plate (length × width × thickness, 50×50×3 mm), (3) copper wire (diameter, 1 mm), (4) electrode (gold-plated copper foil; diameter, 20 mm; thickness, 0.035 mm), (5) guard electrode (gold-plated copper foil; diameter, 22 mm; thickness, 0.035 mm), (6) electrode (gold-plated copper foil; length × width × thickness, 50×50×0.035 mm), (7) epoxy adhesives, (8) epoxy resin (thickness, 1.5–2.5 mm)

Table 5. Typical dielectric cells for measuring epoxy resin systems before and during crosslinking

Type of cell	Electrode				Ref.
	Material	Shape	Number of terminals	Temperature correction[a]	
A. Parallel plate cell					
a. Vertical type	Aluminum	Circle	2[b]	No	71
b. Horizontal type	Unknown[c] (50 mmφ)	Circle	3	No	72
c. Vertical type	Aluminum (4.0 cm²)	Square	2[b]	No	73, 74
d. Horizontal type	Aluminium[d]	Circle	2[b]	No	75
e. Horizontal type	Aluminum	Cylindrical (15 mmφ)	2[b]	Yes	76,77
f. Vertical type	Copper (Gold plated)	Circle (20 mmφ)	3	Yes	10

continued p.154

Table 5. (continued)

Type of cell	Electrode				Ref.
	Material	Shape	Number of terminals	Temperature correction[a]	
B. Microchip cell	Unknown	Comb (0.06×0.10 inch)	2[b]	No	78, 79

[a] Correction of cell for the thermal expansion by temperature
[b] Without a guard electrode
[c] Electrodes are covered with aluminum or blocking polytetrafluoroethylene (PTFE) sheets
[d] Electrodes are covered with aluminum foils

dependent electromagnetic sensor and monitored the curing process of an epoxy graphite laminate in combination with a dynamic viscosity measurement.

4.2.2
Dielectric Relaxation Behavior

The dielectric α-relaxation behavior of an amorphous material is usually expressed with the Debye [83], the Davidson-Cole [84, 85], the Havriliak-Negami [86], or the Kohlrausch-Williams-Watts (KWW) [87] formula. Table 6 summarizes some studies on the dielectric α-relaxation behavior for DGEBA oligomers [10, 88, 92–94] along with an uncured epoxy-curing agent system [89] and a phenoxy resin [12–14]. The Havriliak-Negami or the KWW formula is reported to be applicable to the dielectric relaxation behavior of the DGEBA oligomer before crosslinking [10, 88, 92, 93]. The Cole-Cole plot is often used for the determination of α-relaxation behavior [84]. Typical Cole-Cole plots [10] of two DGEBA oligomers, Epikote 828 ($\bar{M}_w$=388) and Epikote 834 ($\bar{M}_w$=590), are shown in Fig. 7. Each Cole-Cole plot is in a good agreement with the Havriliak-Negami equation that is a comprehensive expression for the Debye and the Davidson-Cole equations.

The Havriliak-Negami equation is given by the following formula [86]:

$$\varepsilon^*(\omega) - \varepsilon_\infty = \frac{\varepsilon_0 - \varepsilon_\infty}{[1 + (i\omega\tau)^{1-\alpha}]^\beta} \tag{4}$$

where $\varepsilon^*(\omega)$ is the complex dielectric constant, ε_∞ is the high-frequency limit to the dielectric constant, ε_0 is the low-frequency limit to the dielectric constant, i is the square root of –1, ω is the angular frequency, τ is the relaxation time, α is the distribution parameter, and β is the skewness parameter. These parameters are associated with the properties of epoxide oligomers. Table 7 shows the best-fit parameters [10] for several DGEBA oligomers obtained according to Eq. (4). Each parameter has been reported to well reflect the nature of the uncured DGEBA oligomer, such as the chain length or the molecular weight distribution [10, 88].

Table 6. Dielectric relaxation behavior of epoxy resin systems before crosslinking

Resin system	Type of relaxation	Temperature dependence of dielectric relaxation time	References
A Dielectric Measurement			
DGEBA[a] oligomers ($-19 \leq T_g \leq 79\ °C$)	KWW[b]	Williams-Landel-Ferry	93
DGEBA oligomers ($-16 \leq T_g \leq 82\ °C$)	Havriliak-Negarni	Williams-Landel-Ferry	10,88
DGEBA oligomer ($T_g = -12.4\ °C$)	non-Debye	Vogel-Fulcher	94
Phenoxy resin[c] ($Tg = 97\ °C$)	KWW and Havriliak-Negami	Vogel-Fulcher	12–14
TGDDM/DDS[d] mixture	Havriliak-Negami	–	89
B. DC Transient Current Measurement			
DGEBA oligomer ($T_g = 45\ °C$)	Havriliak-Negami	Williams-Landel-Ferry	92

[a] DGEBA: diglycidylether of bisphenol-A
[b] KWW Kohlraush-Williams-Watts
[c] Phenoxy resin: PKHH (Union Carbide), $\bar{M}_w = 50700$
[d] TGDDM: tetraglycidyl 4,4'-diaminodiphenylmethane, DDS: 4,4'-diaminodiphenylsulfone

The dielectric α-relaxation in the vicinity of T_g can be measured by the DC transient current method [90]. Takeishi and Mashimo [91] have studied the dielectric α-relaxation of poly(vinyl acetate) (PVAc) near the T_g and computed the dielectric properties in a low-frequency range, from 10^{-6} to 1 Hz, by the Fourier transform of the DC transient current. The PVAc polymer studied was found to have a dielectric α-relaxation process fitting the Havriliak-Negami equation [91]. A DGEBA oligomer, Epikote 1003 ($\bar{M}_w = 2078$), has been investigated by the DC transient current method based on the Fourier transform [92]. The dielectric α-relaxation of the oligomer is governed by the Havriliak-Negami equation at a temperature around the T_g (45 °C).

The α-relaxation behavior for many glass-forming liquids is also described by the KWW relaxation function that is obtained by a transform of the following time domain function to the frequency domain one [87]:

$$\phi(t) = \exp(-(t/\tau)^{\beta'}) \tag{5}$$

where τ is a characteristic relaxation time, and β' is a parameter ranging between zero and unity. Sheppard and Senturia [93] reported that the KWW function provided a better representation of the dielectric relaxation behavior of the DGEBA oligomer after examining the applicability of the KWW and the Davidson-Cole relaxation functions. Alegría et al. [12–14] indicated that the dielectric α-relaxation of a phenoxy resin ($\bar{M}_w = 50700$) was describable by either the KWW or the Havriliak-Negami relaxation functions.

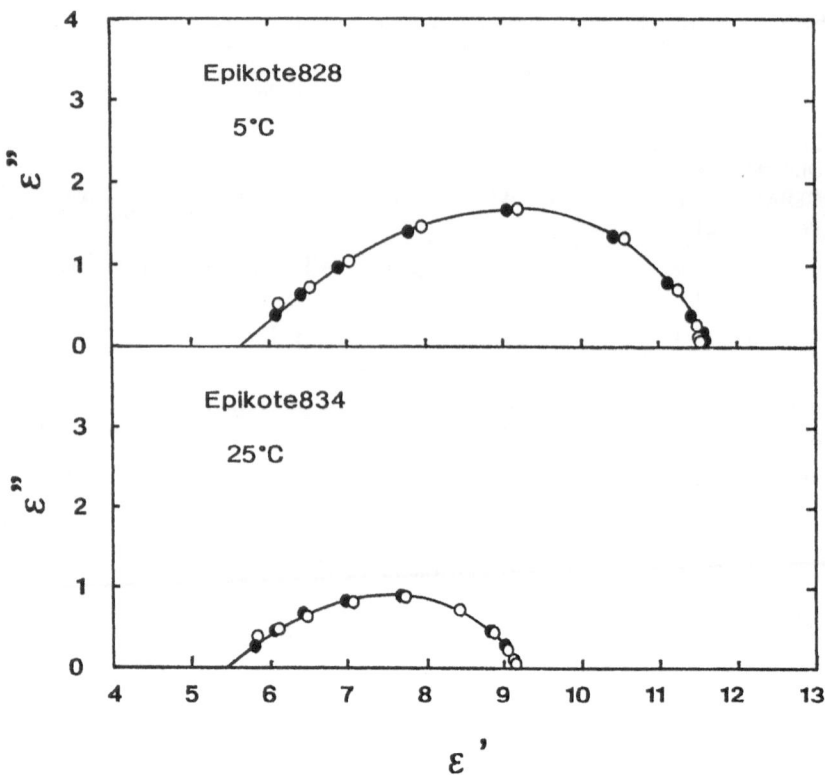

Fig. 7. Cole-Cole arcs for Epikote 828 at 5 °C and Epikote 834 at 25 °C: ○ experimental value; ● calculated value [10]

The temperature dependence of τ of a DGEBA oligomer has been investigated. Two types of descriptions for the temperature dependence of τ are proposed (Table 6): one is the WLF equation [66], the other is the Vogel-Fulcher equation [64, 65]. The WLF equation for the temperature dependence of τ is as follows [66]:

$$\log\frac{\tau(T)}{\tau(T_g)} = \frac{-C_1'(T-T_g)}{C_2'+T-T_g} \tag{6}$$

where T is the temperature ($T_g \leq T \leq T_g + 100$ °C), T_g is the glass transition temperature, and C_1' and C_2' are WLF parameters. The WLF parameters are listed for five DGEBA oligomers ($388 \leq \bar{M}_w \leq 2640$) in Table 4. The other equation, the Vogel-Fulcher one, has the following form [64, 65]:

$$\tau(T) = \tau_0 \exp\frac{DT_0}{T-T_0} \tag{7}$$

Table 7. Parameters of the Havriliak-Negami equation for DGEBA oligomers [10]

DGEBA Oligomers	Temp (°C)	ε_0	ε_∞	$\varepsilon_0-\varepsilon_\infty$	α	β	Relaxation Time (S)
Epikote 828	0	12.00	5.55	6.45	0.24	0.59	3.2×10^{-4}
	5	11.62	5.68	5.94	0.24	0.59	3.0×10^{-5}
	10	11.29	5.77	5.52	0.25	0.60	4.3×10^{-6}
Epikote 834	20	9.43	5.25	4.18	0.32	0.56	2.5×10^{-4}
	25	9.16	5.37	3.79	0.33	0.57	2.7×10^{-5}
	30	8.93	5.38	3.55	0.33	0.57	5.7×10^{-6}
Epikote 1001	50	7.14	5.01	2.13	0.34	0.50	4.2×10^{-4}
	55	7.02	5.06	1.96	0.35	0.51	7.0×10^{-5}
	60	6.91	5.10	1.81	0.35	0.51	1.5×10^{-5}
Epikote 1002	60	6.57	4.84	1.73	0.36	0.50	8.6×10^{-4}
	65	6.44	4.87	1.57	0.36	0.50	1.1×10^{-4}
	70	6.33	4.88	1.45	0.36	0.50	2.4×10^{-5}
Epikote 1004	75	5.83	4.55	1.28	0.37	0.46	5.0×10^{-4}
	80	5.74	4.55	1.19	0.37	0.46	8.3×10^{-5}
	85	5.65	4.55	1.10	0.38	0.47	1.8×10^{-5}

where τ is the dielectric relaxation time at temperature T, and τ_0, D, and T_0 are empirically determined constants. Paluch et al. [94] reported that the Vogel-Fulcher expression was applicable to both the temperature and pressure dependencies of τ for a liquid-type DGEBA oligomer (T_g=–12.4 °C).

4.3
Direct Current Conductivity

The DC conduction in the epoxide oligomer above the T_g is caused by ions (their mobility and concentration) which are always present, to some extent, in commercial products. The DGEBA oligomers produced by a conventional process are considered to have the same kinds of ion. The major ions in the epoxide oligomers are Na^+ and Cl^- due to the manufacturing process. Table 1 shows that the ion concentrations are very low and regarded as similar among the conventional-type DGEBA oligomers having $\bar{M}_w$s from 388 to 2640 [10]. Therefore, the DC conductivity, which is mainly affected by the ionic mobility due to the low concentration of ions in the conventional DGEBA oligomer, is closely related to the segmental mobility of the host oligomer. On the other hand, the advancement-

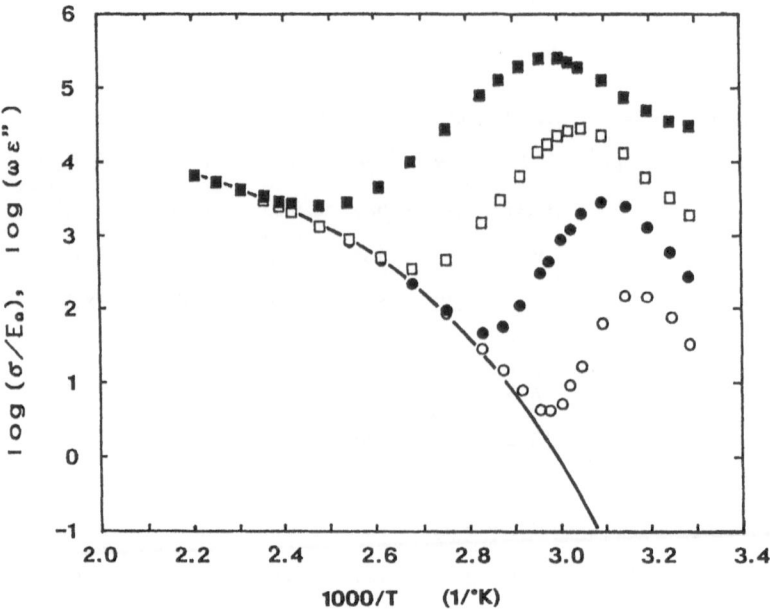

Fig. 8. Temperature dependence of σ/E_0 and $\omega\varepsilon''$ at four frequencies for Epikote 1001: $\bigcirc$ 50 Hz; $\bullet$ 1 kHz; $\square$ 10 kHz; $\blacksquare$ 100 kHz [10]

type DGEBA oligomer contains the residue of the ionic catalyst in concentrations of tens to hundreds ppm. The type of ionic catalyst in the commercially available oligomer is not usually revealed, nor is the concentration of the catalyst. A careful analysis is necessary concerning to the ionic conduction data measured in the commercial advancement-type DGEBA oligomer.

The ionic conductivity of amorphous materials can be obtained not only from the conventional direct current measurement but also from the dielectric loss in the dielectric measurement. The ionic conductivity from the dielectric measurement is based on the following concept:

The dielectric loss ε'', which has a dipolar (ε_d'') and an ionic (ε_i'') component, can be described by Eq. (8) because the ionic term is mainly attributed to the DC conduction [79, 104].

$$\varepsilon''=\varepsilon_d''+\sigma/\omega E_0 \tag{8}$$

where E_0 is the permittivity of free space (8.855×10^{-14} F/cm), σ is the DC conductivity and ω is the angular frequency. Equation (8) can be rearranged in the following form [79, 104]:

$$\omega\varepsilon''=\omega\varepsilon_d''+\sigma/E_0 \tag{9}$$

The term $\omega\varepsilon_d''$ becomes relatively small due to the significant increase in σ at temperatures where active segmental motions occur. If the value of the product

Table 8. Temperature dependence of ionic conductivity for DGEBA oligomers

Relationship	Remarks	Ref.
A. WLF type		
$$\log\frac{\sigma(T)}{\sigma(T_g)}=\frac{C_1{}''(T-T_g)}{C_2{}''+T-T_g}$$	σ : ionic conductivity $C_1{}'', C_2{}''$:WLF parameters T : temperature T_g : glass transition temperature	10, 88
B. Berry & Fox type		
$$\ln\left(\frac{\sigma\eta\zeta_0'}{\zeta_0}\right)=\left(1-\frac{B'}{B}\right)\left[\frac{B}{f_g+\alpha_1(T-T_g)}\right]$$ $\zeta_0=2.49\times10^{-5}$ poise-mole/g $f_g/B=0.025$	σ : ionic conductivity η : melt viscosity B, B' : constants ζ_0 : inherent fraction factor ζ_0' : inherent ion transport factor T_g : glass transition temperature f_g : free volume fraction at T_g T : temperature α_1 : thermal expansion coefficient above T_g	62

$\omega\varepsilon_d{}''$ is negligible, the product $\omega\varepsilon''$ can be a direct measure of the DC conduction.

Figure 8 shows the temperature dependencies of $\omega\varepsilon''$ at four frequencies for Epikote 1001 ($\bar{M}_w=1396$) in comparison with that of σ/E_0 calculated from the data by the DC conduction measurements [10]. A broad peak is observed for each of the four frequencies at low temperatures on the plot of $\omega\varepsilon''$, which is due to the rotational diffusion of the dipole moments. A good agreement is observed between $\omega\varepsilon''$ (plots) and σ/E_0 (a solid curve) at higher temperatures and at lower frequencies in Fig. 8. The dielectric loss ε'' can be used as an indicator of the ionic conduction in the DGEBA oligomer at a fixed frequency at the temperatures where the dipole component is negligible. The ionic conduction from the dielectric loss can be measured in a short period of time and is widely used for the cure analysis of epoxy resin systems [62, 79–82].

The WLF-type equation has been applied to the temperature dependence of σ above the T_g for several polymers by Saito et al. [95]. As for the uncured DGEBA oligomer, the temperature dependence of σ has been also investigated in some of the literature. The WLF-type or Berry & Fox equations have been used for the description of the temperature dependence of σ in the uncured DGEBA oligomer (Table 8) [10, 62, 88]. The two equations are regarded the same in terms of the free-volume concept [62, 67]. The WLF-type equation for σ is expressed in the following form:

$$\log\frac{\sigma(T)}{\sigma(T_g)}=\frac{C_1{}''(T-T_g)}{C_2{}''+T-T_g} \tag{10}$$

where C_1'' and C_2'' are parameters. The best-fit WLF parameters of σ [10] are summarized for five DGEBA oligomers along with those of η and τ in Table 4. Simpson and Bidstrup [62] proposed the other type description, the Berry & Fox equation, for temperature dependencies of some DGEBA oligomers as listed in Table 8.

5
Relationships Between Melt Viscosity, Dielectric Relaxation Time, and DC Conductivity

The melt viscosity (η), the dielectric relaxation time (τ), and the DC conductivity (σ) can be discussed as the same transport phenomena in an amorphous material by considering the movements of three moving units: the chain segment, the dipole, and the ionic charge carrier. Temperature dependencies of these three properties have been studied for DGEBA oligomers as mentioned in a previous section. Figure 9 demonstrates three temperature-dependent curves of η, τ, and σ for a DGEBA oligomer, Epikote 834 ($\bar{M}_w$=590) [61]. Table 9 lists some expressions describing the relationships between the three viscoelastic properties for amorphous materials. In this section, the relationships between two out

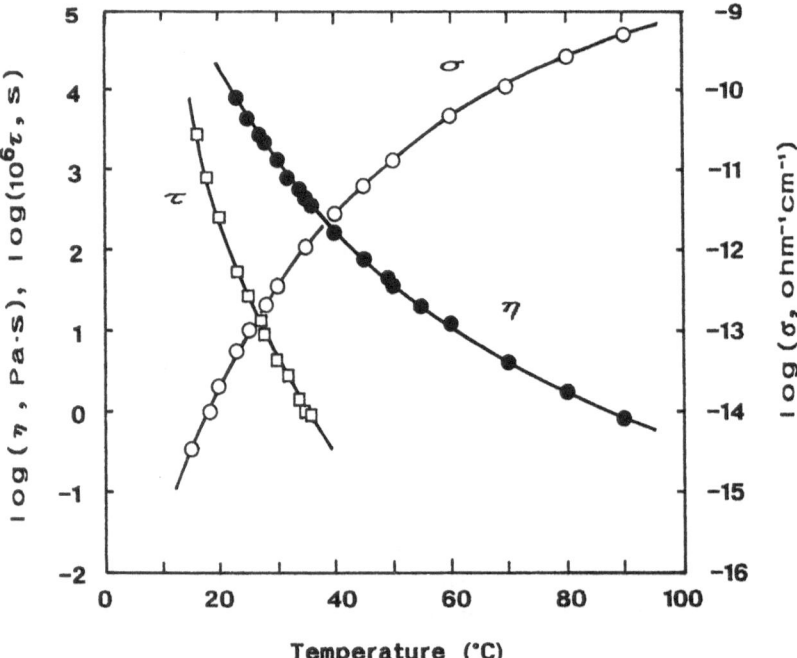

Fig. 9. Temperature dependence of melt viscosity η, DC conductivity σ, and dielectric relaxation time τ for Epikote 834: ● melt viscosity; ○ DC conductivity; □ dielectric relaxation time [61]

Table 9. Relationships between viscosity η, dielectric relaxation time τ, and ionic conductivity σ

Relationship	Material	Ref.
A. η and σ		
$\sigma\eta$ =const	rigid spherical particles (Debye model)	83
$\sigma\eta$ =const	low molecular weight liquids (Walden's rule)	96
$\sigma\eta^v$=const	polystyrene ($\bar{M}_w$=3.2×10^5) solution	98
(1.33≤v≤1.47)	(concentration=0–12 wt% in benzene)	
$\sigma\eta^k$=const	DGEBA oligomers (388≤$\bar{M}$w≤2640)	61
(0.63≤k≤1.12)		
lnσ∝lnη	DGEBA oligomers (346 ≤ $\bar{M}_w$≤14,200)	62
B. σ and τ		
$\sigma\tau^m$=const	poly(vinyl chloride), polycarbonate,	102
(0.18≤m≤1.01)	poly(vinylidene chloride), poly(vinyl acetate),	
	poly(ethylene terephthalate)	
$\sigma\tau^m$=const	DGEBA oligomers (388≤$\bar{M}$w≤9454)	10,88
(0.54≤m≤0.82)		
C. τ and η		
η /τ=const	rigid spherical particles (Debye model)	83
	or polymers	
η/τ^l=const	DGEBA oligomers (388≤$\bar{M}_w$≤2640)	61
(0.73≤l ≤1.06)		

of three properties, η, τ, and σ, are discussed for DGEBA oligomers in comparison with those for polymers or low molecular weight liquids. The methods for predicting viscoelastic properties at T_g are also considered.

5.1
Melt Viscosity and DC Conductivity

It is generally recognized that the ionic conductivity of a low molecular weight liquid is inversely proportional to the bulk viscosity using Stokes' law for the drift of a spherical ion through a viscous medium. The conductivity is the sum of the products of the numbers of ions and their mobilities. The relationship between the DC conductivity (σ) and the melt viscosity (η) is expressed in the following form [79].

$$\sigma= \sum_j N_j Q_j \mu_j =(1/6\pi\eta) \sum_j (N_j Q_j^2/R_j) \tag{11}$$

where N_j is the number/cm^3 of the jth ion, Q_j is the charge on the jth ion, μ_j is the mobility, and R_j is the radius of the jth ion. This inverse proportionality between σ and η, $\sigma \propto 1/\eta$, is known as Walden's rule [96] for low molecular weight liquids. Tajima [97] has applied this relationship to the prediction of viscosity from the real-time conduction measurement during the early curing stage of an epoxy resin system.

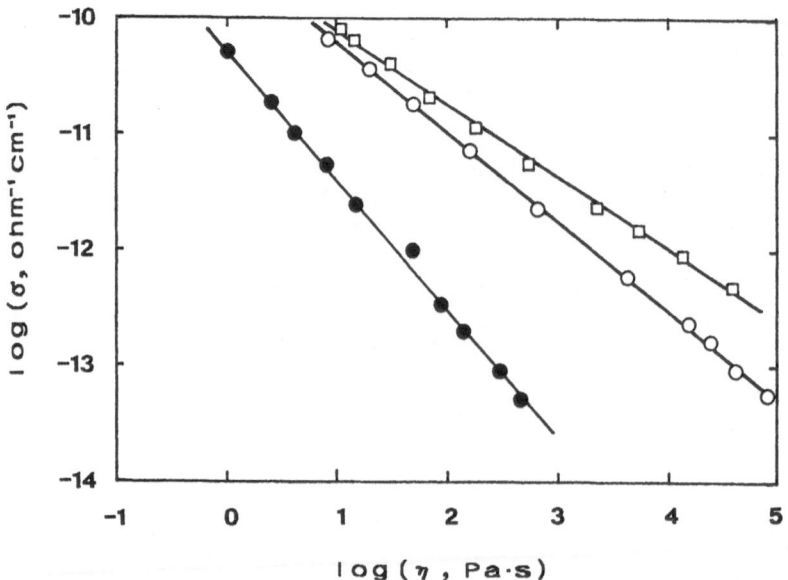

Fig. 10. Relationship between log η and log σ for three DGEBA oligomers: ● Epikote 828; ○ Epikote 1001; ▢ Epikote 1004 [61]

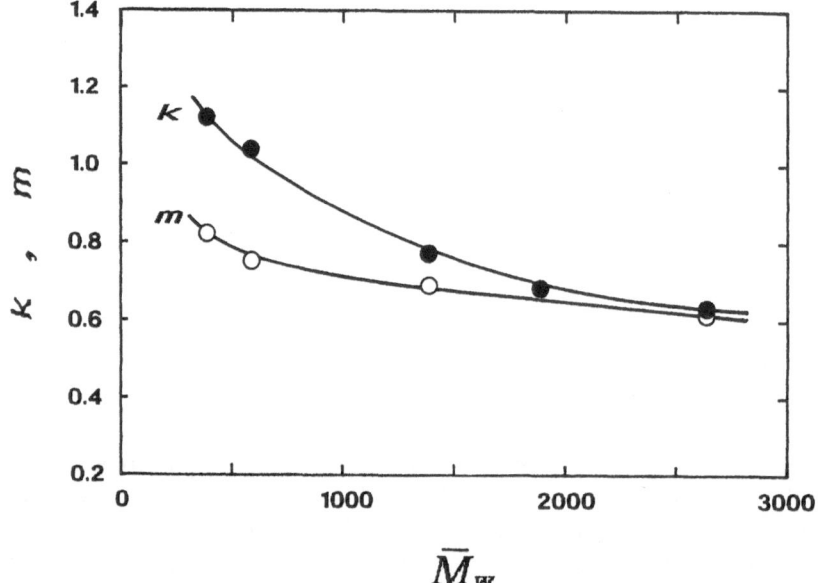

Fig. 11. Plot of exponents k and m vs $\bar{M}_w$ for the DGEBA oligomer: ● k; ○ m [61]

The relationship between σ and η has been examined for the DGEBA oligomer before crosslinking. Log-log plots of σ vs η for three DGEBA oligomers ($388 \leq \bar{M}_w \leq 2640$) are shown in Fig. 10. The linear plots observed between log σ and log η are given by the following formula [61]:

$$\sigma(T)[\eta(T)]^k = const \qquad (12)$$

where $\sigma(T)$ is the DC conductivity at temperature T, $\eta(T)$ is the melt viscosity at T, and k is the exponent ($0.63 \leq k \leq 1.12$). The k value is not always unity. Equation (12) is different from the general understanding, $k=1$ or Walden's rule [96], for low molecular weight liquids. The exponent k is plotted as a function of the $\bar{M}_w$ of the oligomer in Fig. 11 [61]. The exponent k is dependent on the $\bar{M}_w$ of the oligomer and decreases with an increase in the $\bar{M}_w$ of the oligomer. The k value of the low $\bar{M}_w$ oligomer is around unity, which is consistent with Walden's rule for low molecular weight liquids. A higher $\bar{M}_w$ oligomer has a k value lower than unity. The variation in the k value ($0.63 \leq k \leq 1.12$) for these oligomers is probably caused by the mobility difference between the chain segment and the ionic charge carrier [61]. As for high $\bar{M}_w$ polymers, Sazhin and Shuvayev [98] reported that a similar relationship, $\sigma\eta^v = const$ ($1.33 \leq v \leq 1.47$), was applicable to polystyrene ($\bar{M}_w = 3.2 \times 10^5$) solutions. This finding is based on Arrhenius-type temperature dependencies of low viscosity solutions.

The WLF equation is valid over the temperature range from T_g to $T_g + 100$ °C. Equation (12), $\sigma(T)[\eta(T)]^k = const$, is therefore written in the following form at the T_g of the oligomer [99]:

$$\sigma(T_g)[\eta(T_g)]^k = const \qquad (13)$$

The melt viscosity is difficult to measure around the T_g of the epoxide oligomer because the melt viscosity considerably increases near T_g. In contrast to the melt viscosity, the DC conductivity is relatively easy to measure around T_g. Figure 12 demonstrates the relationship between the calculated $\eta(T_g)$ and $\bar{M}_w$ for several DGEBA oligomers ($388 \leq \bar{M}_w \leq 3606$, $-16 \leq T_g \leq 57$ °C) [99]. The $\eta(T_g)$ value of the DGEBA oligomer increases from 10^8 to 10^{12} Pa·s with the increase in $\bar{M}_w$ and appears to saturate at around 10^{12} Pa·s for high $\bar{M}_w$ oligomers ($\bar{M}_w \geq 3000$). The melt viscosity at the T_g is around 10^{12} Pa·s for many polymers [100]. The behavior of $\eta(T_g)$ of the oligomer in a higher $\bar{M}_w$ region ($\bar{M}_w \geq 3000$) agrees with the concept generally accepted for polymers. The glass transition of a polymer is regarded as the state in which the free-volume fraction is equal to ~ 0.025 [100, 101]. This means that the melt viscosity is not always a constant at the T_g and essentially depends on the nature of the polymer. A polymer has some degree of molecular entanglement. The differences in the melt viscosity at T_g between different polymers become smaller for long chain polymers due to entanglements. A low $\bar{M}_w$ DGEBA oligomer, based on a rigid aromatic structure, is thought to have little molecular entanglement. The nature of the low $\bar{M}_w$ DGEBA oligomer will be reflected in the $\eta(T_g)$ value. This is considered to be the reason for the $\bar{M}_w$ dependency of $\eta(T_g)$ in the low $\bar{M}_w$ region of the oligomer [99]. Similar

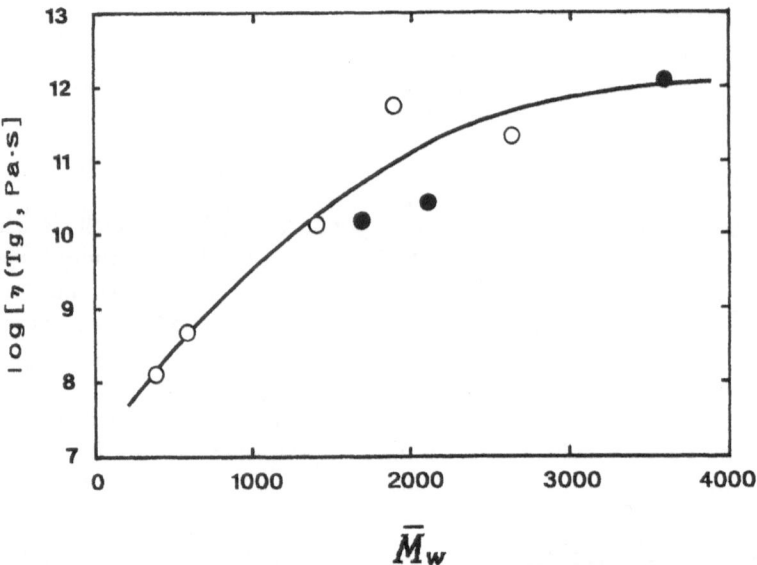

Fig. 12. Relationship between melt viscosity at T_g, $\eta(T_g)$, and $\bar{M}_w$ for DGEBA oligomers: ○ conventional-type oligomer; ● advancement-type oligomer [99]

$\eta(T_g)$ values (from 1.63×10^8 to 3.04×10^{11} Pa·s) have also been reported for some DGEBA oligomers ($-15 \leq T_g \leq 79$ °C) by Utracki and Ghijsels [60].

5.2
DC Conductivity and Dielectric Relaxation Time

Sasabe and Saito [102] reported that the following empirical equation [Eq. (14)] was observed between τ and σ for some amorphous polymers: poly(vinyl chloride), poly(vinylidene chloride), poly(vinyl acetate), amorphous poly(ethylene terephthalate), and polycarbonate.

$$\sigma(T)[\tau(T)]^m = \text{const} \tag{14}$$

where T is temperature and m is an exponent. The m-value for these polymers varies from 0.18 to 1.01 and decreases with increasing T_g. Sasabe and Saito [102] explained the $\bar{M}_w$ dependence of m by the mobility difference between the dipole (or the chain segment) and the ion. The relationship between τ and σ has also been examined for DGEBA oligomers. Figure 13 demonstrates the log-log plots of τ vs σ for five DGEBA oligomers [10]. Linear plots are observed between $\log \tau$ and $\log \sigma$, which indicates that for the DGEBA oligomers a similar relationship between τ and σ, $\sigma(T)[\tau(T)]^m = \text{const}$, is valid as for other polymers. The $\bar{M}_w$ dependence of m [10] is plotted along with k in Fig. 11. The m-values decrease from 0.82 to 0.62 with increasing $\bar{M}_w$ of the oligomer, similarly to the polymers previously mentioned [10, 102].

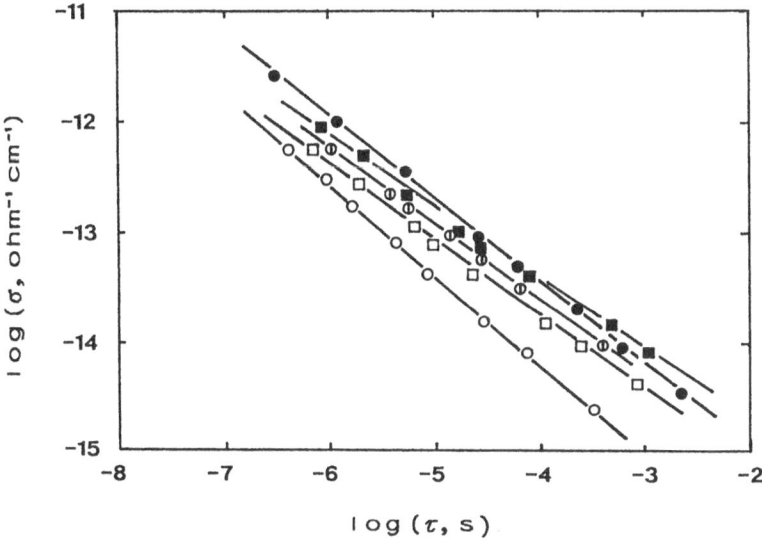

Fig. 13. Relationship between log τ and log σ for five DGEBA oligomers: ○ Epikote 828; ● Epikote 834; ⊕ Epikote 1001; ⊔ Epikote 1002; ■ Epikote 1004 [10]

The relationship between τ and σ can also be used to predict the dielectric relaxation time at T_g, $\tau(T_g)$. Equation (14) provides the following relationship between $\tau(T_g)$ and $\sigma(T_g)$ [88, 92]:

$$\sigma(T_g)[\tau(T_g)]^m = const \qquad (15)$$

The $\tau(T_g)$ is difficult to measure directly by conventional dielectric bridge measurement, but the $\tau(T_g)$ value is calculated from $\sigma(T_g)$ through Eq. (15). The $\tau(T_g)$ values have been reported to be from 12 to 69 s for some DGEBA oligomers ($1396 \leq \bar{M}_w \leq 9454$, $30 \leq T_g \leq 82$ °C) [88]. The $\tau(T_g)$ value has no $\bar{M}_w$ dependency for the oligomers studied. It is the same within experimental error taking into account the accuracy of the measurement of T_g (±1 °C).

5.3
Dielectric Relaxation Time and Melt Viscosity

The relaxation time is generally believed to be directly proportional to the melt viscosity for a simple liquid whose dielectric α-relaxation is governed by the Debye model. The dielectric relaxation time of a simple liquid is expressed by the following relationship [83, 103, 104]:

$$\tau = \zeta/2k_B T \qquad (16)$$

where ζ is the friction constant, k_B is the Boltzmann constant, and T is the absolute temperature. By treating the molecule dipole as a sphere of radius a rotating

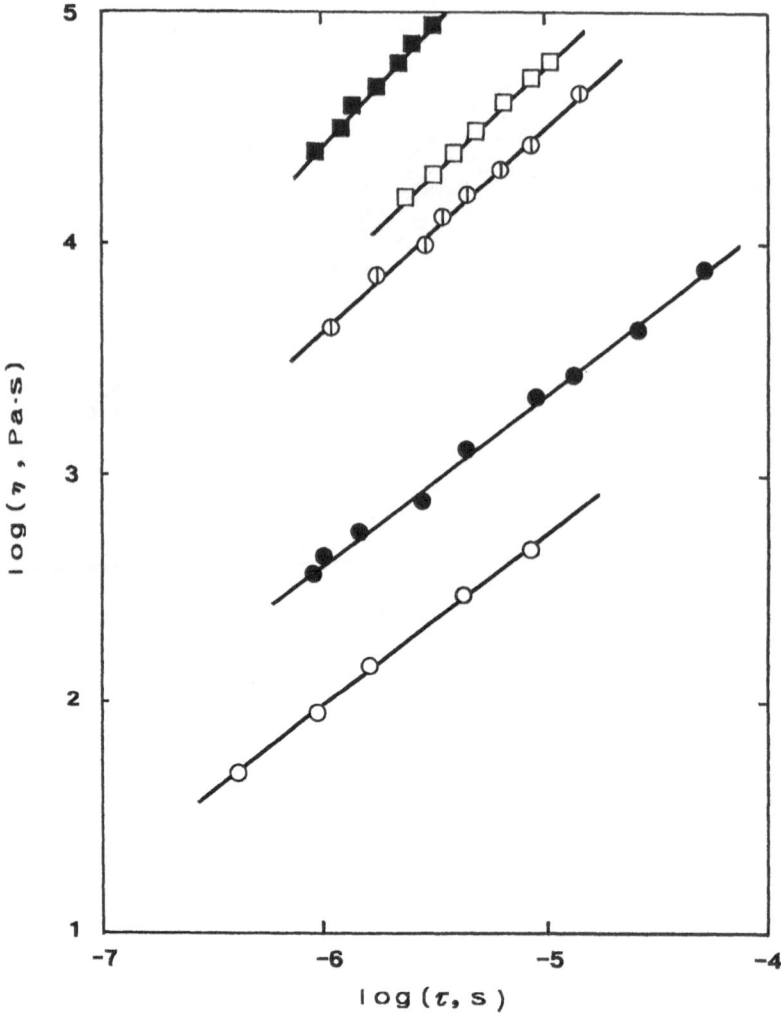

Fig. 14. Relationship between log η and log τ for five DGEBA oligomers: ○ Epikote 828; ●
Epikote 834; ◑ Epikote 1001; ❑ Epikote 1002; ■ Epikote 1004 [61]

in a continuous viscous medium of macroscopic viscosity η, Eq. (16) can also be
expressed in the following form [83,103,104]:

$$\tau=4\pi\eta a^3/k_B T \tag{17}$$

This equation means that the relaxation time is directly proportional to the
melt viscosity for simple liquids. Many polymers also exhibit a direct proportionality between τ and η similarly as simple liquids [100].

The dielectric α-relaxation of the DGEBA oligomer is not fitted to the Debye
model that is applicable to simple liquids. The correlation between η and τ of the

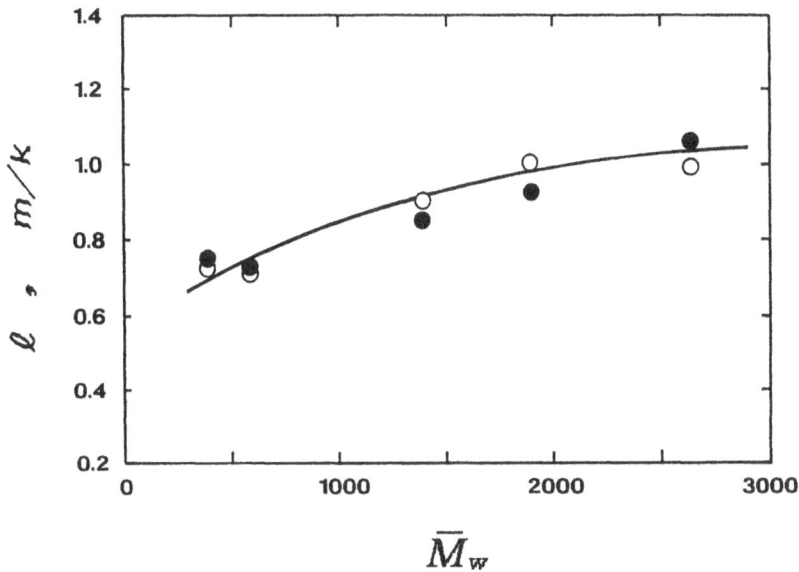

Fig. 15. Plot of exponents l and m/k vs $\bar{M}_w$ for the DGEBA oligomer: ● l; ○ m/k [61]

DGEBA oligomer was examined considering the difference from that of a simple liquid or a polymer. The relationship between η and τ of the DGEBA oligomer is numerically derived from the two relationships, $\sigma(T)[\eta(T)]^k=$const and $\sigma(T)[\tau(T)]^m=$const, already obtained. The relationship derived is expressed in the following form [61]:

$$\eta(T)/[\tau(T)]^{m/k}=\text{const} \tag{18}$$

where m/k is the exponent. The value of m/k is between 0.72 and 1.0 for the DGEBA oligomers having $\bar{M}_w$ between 388 and 2640, which implies that the relaxation time is not always proportional to the melt viscosity of the oligomers. As shown in Fig. 9, a direct comparison of the melt viscosity with the dielectric relaxation time is possible in a relatively narrow range of temperatures for the DGEBA oligomer. Figure 14 demonstrates log-log plots of the melt viscosity vs dielectric relaxation time for five DGEBA oligomers. The relationship between $\log \eta$ and $\log \tau$ for each oligomer is also linear, which gives the following empirical relation between η and τ [61]:

$$\eta(T)/[\tau(T)]^{l}=\text{const} \tag{19}$$

where l is the exponent ($0.73 \leq l \leq 1.06$). This equation is of the same type as Eq. (18). The exponent l is plotted as a function of the $\bar{M}_w$ of the DGEBA oligomer in comparison with the exponent m/k in Fig. 15. A good agreement is observed between the two exponents; the experimentally obtained l ($0.73 \leq l \leq 1.06$) and the calculated m/k ($0.72 \leq m/k \leq 1.00$) [61].

The dielectric relaxation time is regarded to be directly proportional to the melt viscosity when the exponent l in Eq. (19) is near unity. Figure 15 shows that the dielectric relaxation time is not directly proportional to the melt viscosity for low $\bar{M}_w$ oligomers ($\bar{M}_w$<2000) as indicated by the l values lower than unity [61]. Generally, the dielectric relaxation time is directly proportional to the melt viscosity for a simple liquid the dielectric α-relaxation of which is governed by the Debye model [83]. This also implies that the direct proportionality between η and τ is not always valid for the compound having the non-Debye-type relaxation [90]. The low $\bar{M}_w$ DGEBA oligomer is considered as one such example [61]. In contrast to the low $\bar{M}_w$ oligomer, a high $\bar{M}_w$ DGEBA oligomer exhibits the direct proportionality between η and τ (Fig. 15), which agrees with the generally accepted behavior of a higher molecular weight polymer. The relationship between η and τ will be discussed below from the standpoint of the mobility difference between the chain segment and the dipole.

6
Behavior of Williams-Landel-Ferry Parameters

The same WLF-type equations are applicable to the temperature dependencies of the melt viscosity, the dielectric relaxation time, and the DC conductivity for uncured DGEBA oligomers. The WLF C_1 and C_2 parameters [10] are plotted as a function of the $\bar{M}_w$ for the oligomer in Figs. 16 and 17. The WLF parameters have molecular weight dependencies for the DGEBA oligomers. Three C_1 parameters, C_1, C_1', and C_1'', have different values for the same oligomer, while the other parameters, C_2, C_2', and C_2'', have similar values. The relationships between these WLF parameters for the DGEBA oligomers are summarized in the following form [10]:

$$C_1 \neq C_1' \neq C_1'' \tag{20}$$

$$C_2 \approx C_2' \approx C_2'' \tag{21}$$

The behavior of the WLF parameters can be explained using the free-volume concept. The two WLF parameters are described in the following form according to the free-volume theory developed by Cohen and Turnbull [105,106]:

$$C_1 = \gamma \upsilon^* / (2.303 f_g \bar{\upsilon}_m) \text{ and } C_2 = f_g / \alpha_f \tag{22}$$

where C_1, C_2 are the WLF parameters, γ is a numerical factor introduced to correct for the overlap of the free-volume, υ^* is the critical volume large enough to permit a molecule to jump in after the displacement, $\bar{\upsilon}_m$ is the average molecular volume, f_g is the free-volume fraction at the T_g, and α_f is the thermal expansion coefficient of the free-volume. The C_1 parameter is determined not only by the characteristics of the oligomer (a host oligomer) matrix but also by the size of the moving unit expressed by the critical volume, υ^*, for the transport of the oligomer segment, dipole, or ionic charge carrier [Eq. (22)]. Therefore, three C_1 parameters, C_1, C_1', and C_1'', take essentially different values for the same oli-

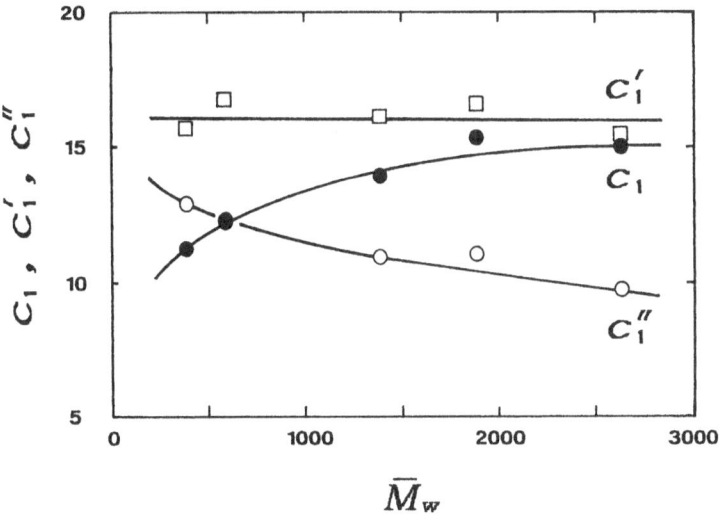

Fig. 16. Plot of WLF C_1 parameters for η, τ, and σ, vs $\bar{M}_w$ for the DGEBA oligomer: ● C_1 for η; ◻ C_1' for τ; ○ C_1'' for σ [61]

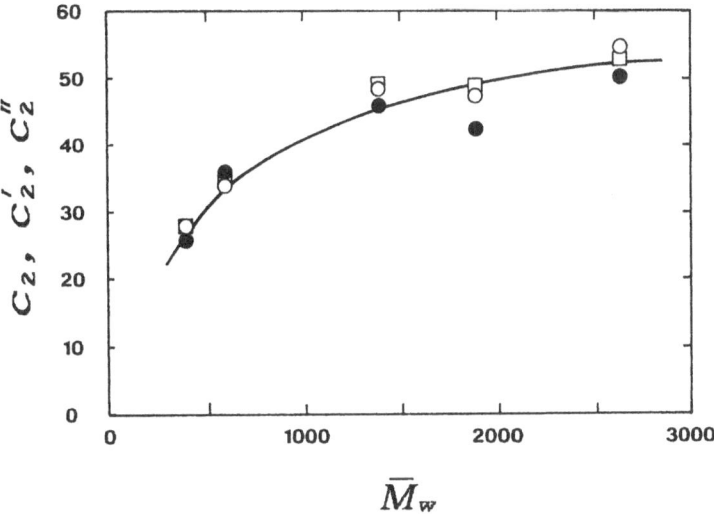

Fig. 17. Plot of WLF C_2 parameters for η, τ, and σ, vs $\bar{M}_w$ for the DGEBA oligomer: ● C_2 for η; ◻ C_2' for τ; ○ C_2'' for σ [61]

gomer as shown in Fig. 16. The other WLF parameter, C_2, reflects the characteristics of an oligomer matrix or a host oligomer matrix only, which leads to the C_2 parameters, C_2, C_2' and C_2'' taking similar values for the same oligomer as demonstrated in Fig. 17.

Table 10. Comparison of exponents and the ratios between the WLF C_1 parameters for DGEBA oligomers [61]

DGEBA Oligomer	k	C_1''/C_1	l	C_1/C_1'	m	C_1''/C_1'	m/k
Epikote 828	1.12	1.15	0.75	0.72	0.82	0.82	0.73
Epikote 834	1.04	1.00	0.73	0.73	0.75	0.73	0.72
Epikote 1001	0.77	0.79	0.85	0.86	0.69	0.68	0.90
Epikote 1002	0.68	0.72	0.92	0.93	0.68	0.67	1.00
Epikote 1004	0.63	0.65	1.06	0.97	0.62	0.63	0.99

Table 11. Relationships between WLF parameters and viscoelastic properties for DGEBA oligomers

Relationship	References
A. Temperature dependence	
a. Melt viscosity (η)	
$\log\dfrac{\eta(T)}{\eta(T_g)} = \dfrac{-C_1(T-T_g)}{C_2+T-T_g}$	61, 99
b. Dielectric relaxation (τ)	
$\log\dfrac{\tau(T)}{\tau(T_g)} = \dfrac{-C_1'(T-T_g)}{C_2'+T-T_g}$	10, 88
c. Direct current conductivity (σ)	
$\log\dfrac{\sigma(T)}{\sigma(T_g)} = \dfrac{C_1''(T-T_g)}{C_2''+T-T_g}$	10, 99
B. Relationship between WLF parameters	
$C_1 \neq C_1' \neq C_1''$, $C_2 \approx C_2' \approx C_2''$	10, 61
C. Relationships between η, τ, and σ	
a. η and σ	
$\sigma(T)[\eta(T)]^k = \text{const}$ $\quad\left(\begin{array}{l} k \approx 1^{\mathrm{a}} \text{ for low } \bar{M}_w \text{ oligomers} \\ k<1 \text{ for high } \bar{M}_w \text{ oligomers} \end{array}\right)$ $k \approx C_1''/C_1$	61, 99
b. σ and τ	
$\sigma(T)[\tau(T)]^m = \text{const}\ (m<1)$ $m \approx C_1''/C_1'$	10, 61, 88
c. τ and η	
$\eta(T)/[\tau(T)]^l = \text{const}$ $\quad\left(\begin{array}{l} l<1 \text{ for low } \bar{M}_w \text{ oligomers} \\ l \approx 1^{\mathrm{b}} \text{ for high } \bar{M}_w \text{ oligomers} \end{array}\right)$ $l \approx C_1''/C_1'$	61

[a] $k \approx 1$: σ is inversely proportional to η (Walden's rule)
[b] $l \approx 1$: τ is directly proportional to η

Sasabe and Saito [102] have found that the exponent m in the relationship, $\sigma(T)[\tau(T)]^m$=const, corresponds to the ratio of two WLF C_1 parameters, C_1''/C_1', for some polymers. The ratios of two C_1 parameters, C_1''/C_1, C_1/C_1', and C_1''/C_1', have been investigated for DGEBA oligomers. Table 10 lists the ratios of two C_1 parameters along with three exponents, k, l, and m, for five DGEBA oligomers [61]. By comparing the ratios of two C_1 parameters and the exponents, the following relationships are experimentally obtained [61]:

$$k \approx C_1''/C_1, \; l \approx C_1/C_1', \text{ and } m \approx C_1''/C_1' \qquad (23)$$

The third relationship, $m \approx C_1''/C_1'$, agrees with Sasabe's work for some polymers [102].

These relationships can also be numerically derived. Equation (12), $\sigma(T)[\eta(T)]^k$=const ($T_g \leq T \leq T_g+100\ °C$), can be written in the following form when Eq.(12) is valid at T_g [99]:

$$\sigma(T)[\eta(T)]^k = \sigma(T_g)[\eta(T_g)]^k \qquad (24)$$

$$[\eta(T)/\eta(T_g)]^k = \sigma(T_g)/\sigma(T) \qquad (25)$$

Therefore, k is described as

$$k = -\log[\sigma(T)/\sigma(T_g)]/\log[\eta(T)/\eta(T_g)] \qquad (26)$$

The WLF C_2 parameters for η and σ are expected to have the same value for the same oligomer or host oligomer according to the relationship, $C_2=f_g/\alpha_f$. When the C_2'' parameter for σ has the same value as C_2 for η, dividing one WLF equation [Eq. (10)] for σ by another [Eq. (6)] for η gives the following simple relationship:

$$\log[\sigma(T)/\sigma(T_g)]/\log[\eta(T)/\eta(T_g)] = -C_1''/C_1 \qquad (27)$$

Equation (27) has the same form as Eq. (26). The k value, therefore, is considered to correspond to C_1''/C_1 and is one of the measures describing the relationship between η and σ [99]. Two other relations, $l \approx C_1/C_1'$ and $m \approx C_1''C_1'$, can be numerically derived in the same manner. The behavior of the WLF parameters and the relationships between η, τ, and σ are summarized for the DGEBA oligomer in Table 11.

7
Free-Volume in Epoxide Oligomers

The macroscopic free-volume is an important parameter closely related to the rheological behavior of an amorphous material. It is generally accepted that the glass transition is regarded as an iso-free-volume state, and the free-volume fraction (f_g) at the T_g is around 0.025 for many monomers and polymers. The WLF method has been widely used for determining the macroscopic f_g value. On the other hand, the microscopic analysis of the free-volume has also been utilized for some epoxy resin systems. Positron annihilation spectroscopy (PAS) [107–112] provides useful information on the microstructural properties of free-volume, such as free-volume hole sizes and distributions. The PAS method has been applied to epoxy resin systems during and after crosslinking. Deng et al. [108–110] attempted to calculate the fractional free volume in the crosslinked epoxy network from the ortho-positronium lifetime and corresponding intensity. Wang et al. [111] estimated the fractional free-volume of a crosslinked epoxy resin from the PAS measurement on the assumption that the f_g is 0.025.

The free-volume generally has a size distribution. The macroscopic average free-volume, however, is still an important parameter associated with the rheological behavior of epoxy resin systems. In this section, the thermal expansion behavior and the macroscopic average free-volume based on the WLF method are discussed for DGEBA oligomers before crosslinking in comparison with crosslinked DGEBA networks.

7.1
Thermal Expansion Behavior

In Fig. 18, the volume-expansion behavior is demonstrated for the four uncured DGEBA oligomers, Epikote 1001, 1002, 1003, and 1004 ($1396 \leq \bar{M}_w \leq 2640$) [113]. Thermal expansion coefficients above and below T_g, α_l and α_g, are from 5.34×10^{-4} to $5.54 \times 10^{-4}\,°C^{-1}$ and from 1.30×10^{-4} to $1.50 \times 10^{-4}\,°C^{-1}$, respectively, for these DGEBA oligomers (Table 12). The value of $\alpha_l - \alpha_g$ decreases with an increase in the T_g of the oligomer. The thermal-expansion behavior of the uncured DGEBA oligomer is compared with that of the polymer using the following relationship proposed for amorphous polymers by Simha and Boyer [114, 115]:

$$(\alpha_l - \alpha_g)\, T_g \approx const \approx 0.113 \tag{28}$$

where T_g is in K. The product, $(\alpha_l - \alpha_g)\, T_g$, of the DGEBA oligomer ranges from 0.124 to 0.129 and is similar among the oligomers ($1396 \leq \bar{M}_w \leq 2640$) investigated. Several aromatic backbone polymers, such as polystyrene, have the $(\alpha_l - \alpha_g) T_g$ values over the range from 0.101 to 0.179 (the average value: 0.128) [115]. These four oligomers ($1396 \leq \bar{M}_w \leq 2640$) are thought to exhibit a behavior similar to high $\bar{M}_w$ polymers with respect to the volume expansion.

Table 12 lists the α_l and α_g values for DGEBA oligomers [113], an uncured DGEBA-aromatic amine mixture [116], and cured DGEBA networks [36,116–121]. Plazek and Choy [116] measured the volume-expansion behavior of an uncured mixture of Epon 828 (a low $\bar{M}_w$ DGEBA oligomer equivalent to Epikote 828) and 4,4'-diaminodiphenylsulfone (DDS). The α_l value ($6.66 \times 10^{-4}\,°C^{-1}$) of Epikote 828 is close to that ($6.6 \times 10^{-4}\,°C^{-1}$) of the Epon 828-DDS mixture before crosslinking. This is understandable because both Epikote 828 and the Epon 828-DDS mixture have similar molecular weights and are based on similar aromatic backbone structures. An uncured DGEBA oligomer is thought to have an α_l value greater than that of the cured DGEBA network because the crosslinked

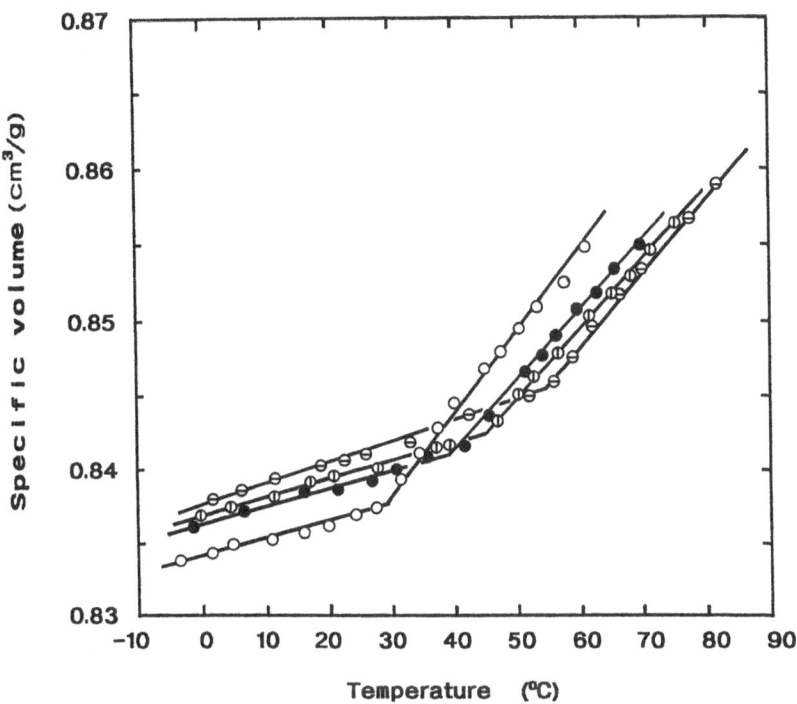

Fig. 18. Specific volume as a function of temperature for DGEBA oligomers: oligomers: ○ Epikote 1001; ● Epikote 1002; ① Epikote 1003; ⊖ Epikote 1004 [113]

Table 12. Thermal expansion behavior and fractional free volume at T_g for epoxy resin systems before and after crosslinking

Epoxy systems	T_g (°C)	$\alpha_1^a \times 10^4$ (°C^{-1})	$\alpha_1^b \times 10^4$ (°C^{-1})	$(\alpha_1-\alpha_g) \times 10^4$ (°C^{-1})	f_g	Ref.
A. Before crosslinking						
Epikote 828	-16	6.66^c	–	–	–	–
Epikote 834	0	6.03^c	–	–	–	–
Epikote 1001	30	5.54	1.30	4.24	0.021^d	113
Epikote 1002	39	5.34	1.35	3.99	0.019^d	113
Epikote 1003	45	5.34	1.36	3.98	0.020^d	113
Epikote 1004	54	5.44	1.50	3.94	0.022^d	113
Epon 828/DDS^e mixture	11	6.6	–	–	–	116
B. After crosslinking^f						
Epon 828/DSA/DMP30^g	62	5.3	2.6	2.7	–	117
Epon 828/DSA/DMP30	37	5.6	2.4	3.2	–	117
Epon 828/DDS	212	5.5	2.8	2.7	–	116
Epon 1001F/DDS	132	4.3	1.8	2.6	–	116

cont. p.174

Table 12. (continued)

Epoxy systems	T_g (°C)	$\alpha_l{}^a \times 10^4$ (°C^{-1})	$\alpha_l{}^b \times 10^4$ (°C^{-1})	$(\alpha_l-\alpha_g) \times 10^4$ (°C^{-1})	f_g	Ref.
B. After crosslinking						
Epon 1004F/DDS	113	5.3	2.2	3.1	–	116
Epon 1007F/DDS	105	5.8	2.6	3.2	–	116
Epon 828/MDA^h	134	–	–	3.14	0.026^d	118
Epon 828/MTHPA^i	119	–	–	3.84	0.025^d	118
Epon 828/2E4Mz^j	75	–	–	3.72	0.021^d	118
GY 250/mPDA^k	136	3.50	1.56	1.94	–	119
Epikote 828/DMP30	130	6.02	3.22	2.80	–	120
Epikote 828/EDA^l	135	6.60	2.60	4.00	–	121
Epikote 828/TMI)A^m	126	6.91	2.60	4.31	–	121
Epikote 828/HMDA^n	110	6.07	2.72	3.35	–	121
Epikote 828/DMDA^o	75	6.52	2.81	3.71	–	121
Epon 828/DEA^p	ca. 60	5.50	1.85	3.65	–	36

^{a} Thermal expansion coefficient above T_g
^{b} Thermal expansion coefficient below T_g
^{c} Author's unpublished data
^{d} f_g obtained from ionic conduction measurement
^{e} DDS: 4,4'-diaminodiphenylsulfone
^{f} Stoichiometric compositions except catalytic curing agents
^{g} DSA: dodecenyl succinic anhydride, DMP30: 2,4,6-tris (dimethlaminomethyl) phenol
^{h} MDA: methylene dianiline
^{i} MTHPA: methyl tetrahydrophtalic acidanhydride
^{j} 2E4Mz: 2-ethly 4-methylimidazole
^{k} GY250: DGEBA (epoxy equivalent =180–190) mPDA: metaphenylene diamine
^{l} EDA: ethylenediamine
^{m} TMDA: tetramethlenediamine
^{n} HMDA: hexamethlenediamine
^{o} DMDA: dodecamethylenediamine
^{p} DEA: diethanolamine

network restricts, to some extent, the thermal expansion above the T_g as suggested Bellenger et al. [122]. A relatively large α_l value is observed for the DGEBA oligomer before crosslinking when comparing uncured and cured DGEBA systems based on the similar $\bar{M}_w$ oligomers (Table 12). As for thermal expansion below the T_g, the α_g value (1.30×10^{-4} °C^{-1}) of uncured Epikote 1001 is smaller than that (1.8×10^{-4} °C^{-1}) of the crosslinked Epon 1001F–DDS network. A similar behavior is observed between Epikote 1004 before crosslinking and the Epon 1004F-DDS network.

7.2
Free-Volume in Epoxide Oligomer Before Crosslinking

The free-volume fraction (f_g) at T_g can be obtained from either the WLF C_1 or C_2 parameter. The WLF C_1 parameter, $C_1=\gamma\upsilon^*/(2.303f_g\bar{\upsilon}_m)$, has two unknown factors, γ and υ^*. The f_g is then usually determined from the C_1 parameter using the following simple relationship for many polymers:

$$f_g = 1/(2.303C_1) \tag{29}$$

This relationship is based on the assumption, $\gamma\upsilon^*/\bar{\upsilon}_m = 1$, for the mobility of a moving unit. The C_1 parameter of the uncured DGEBA oligomer varies depending on the moving unit measured as previously discussed. Therefore, the other WLF parameter, C_2, is utilized for the f_g determination of the uncured DGEBA oligomer through the following relationship:

$$f_g = C_2\alpha_f \tag{30}$$

where α_f is the thermal expansion coefficient of the free-volume. The α_f value is usually approximated by the value of $\alpha_l - \alpha_g$ because the thermal expansion coefficient of the free-volume is difficult to measure directly. Three WLF C_2 parameters are obtained from the dielectric relaxation time, the melt viscosity, and the DC conduction measurements for the same DGEBA oligomer. The WLF C_2 parameters are reported to be similar for the three moving units in the same DGEBA oligomer. Three kinds of f_g values, therefore, are also similar for the same DGEBA oligomer. Table 12 lists the f_g values ($0.019 \leq f_g \leq 0.022$) based on the ionic conduction for four DGEBA oligomers, Epikote 1001, 1002, 1003, and 1004 ($1396 \leq \bar{M}_w \leq 2640$). These f_g values are smaller than the universal f_g value (0.025).

7.3
Free-Volume in Crosslinked Network

The f_g value of the crosslinked system is also estimated from either the WLF C_1 or C_2 parameter. Gupta and Brahatheeswaran [119] obtained the C_1 parameter from the torsional braid analysis (TBA) and calculated the free-volume of the crosslinked DGEBA system using the relationship, $f_g = 1/(2.303C_1)$. The f_g value is reported to be from 0.017 to 0.034 for several DGEBA-aromatic amine networks having different mixing ratios. Ogata et al. [123] applied the same relationship between f_g and C_1 to the calculation of free-volume in the epoxy resin network. The f_g values obtained vary between 0.032 and 0.048 for the epoxy resin systems cured with a novolac-type phenol resin in combination with different accelerators. These results suggest that further study is necessary to confirm the validity of the assumption, $\gamma\upsilon^*/\bar{\upsilon}_m = 1$, in the f_g prediction for the crosslinked epoxy resin network.

The other parameter, C_2, has been used for the f_g determination of the crosslinked DGEBA network. Miyamoto and Shibayama [118,124] calculated the fractional free-volume of the DGEBA network using the C_2 parameter based on the ionic conduction. The DC conduction measurement is the method investigating the same kind of moving unit, the ionic charge carrier, both in the uncrosslinked oligomer and in the crosslinked network. Table 12 summarizes the f_g values that are obtained from the DC conduction measurement for three crosslinked DGEBA systems. The f_g value of the crosslinked DGEBA network is between 0.021 and 0.026 that is similar or slightly greater than that ($0.019 \leq f_g \leq 0.022$) of the DGEBA oligomer before crosslinking [113, 118]. There is a reason supporting a larger f_g value in the crosslinked DGEBA network. The molecular packing and the hydrogen bonding may have influence on the free-volume development in the crosslinked system [125–128]. Gupta and Bra-hatheeswaran [127,128] suggest that the crosslink site based on the bulky bi-sphenol-A structure can be a source of free-volume because the crosslink site does not always provide a suitable environment for close packing of rigid mole-cules.

8
Mobility Difference Between Chain Segment, Dipole, and Ionic Charge Carrier

8.1
Critical Volume for Transport of Moving Unit

The critical volume (υ^*) for the transport of a moving unit is a measure of the mobility of the chain segment, the dipole, or the ionic charge carrier. The υ^* val-ue is difficult to obtain through Eq. (22), $C_1 = \gamma \upsilon^*/(2.303 f_g \bar{\upsilon}_m)$, because γ, a cor-rection factor for the overlap of free-volume, is still undetermined for the DGE-BA oligomer. Cohen and Turnbull [105, 106] reported that γ took a value from 0.5 to unity for some simple liquids. The product $\gamma \upsilon^*$ can be used as one of the indicators for the mobility of each moving unit in the DGEBA oligomer. Equa-tion (22) shows that the $\gamma \upsilon^*$ value is given by [113]:

$$\gamma \upsilon^* = 2.303 f_g \bar{\upsilon}_m C_1 \tag{31}$$

The $\gamma \upsilon^*$ value can be obtained for each moving unit: the chain segment, the di-pole, or the ionic charge carrier. The three $\gamma \upsilon^*$ values, $\gamma \upsilon^*_\eta$, $\gamma \upsilon^*_\tau$, and $\gamma \upsilon^*_\sigma$, are plot-ted as a function of the $\bar{M}_w$ for the four conventional-type DGEBA oligomers, Epikote 1001, 1002, 1003, and 1004 ($1396 \leq \bar{M}_w \leq 2640$), in Fig. 19 [113]. Although γ is an undetermined factor, the $\bar{M}_w$ dependence of $\gamma \upsilon^*$ provides useful informa-tion on the mobility of each moving unit. The $\gamma \upsilon^*_\eta$ value for the segment mobility increases with an increase in the $\bar{M}_w$ of the oligomer. If the difference in the cor-rection factor γ is small between the same kinds of oligomers ($1396 \leq \bar{M}_w \leq 2640$), the $\bar{M}_w$ dependence of $\gamma \upsilon^*_\eta$ for the segment mobility in Fig. 19 is considered reasonable; a lower $\bar{M}_w$ oligomer needs a smaller critical volume for the trans-port of a chain segment. The behavior of $\gamma \upsilon^*_\sigma$ for the ion mobility is also ex-plained by considering the size of the moving unit. The four DGEBA oligomers are produced from the same raw materials by the same production process and are thought to have the same kinds of ion [10]. The major ions in the epoxide ol-igomers are Na^+ and Cl^- due to the manufacturing process. The concentrations of Na^+ and Cl^- are very low and similar among these oligomers [10]. The size of the ion is much smaller than that of the oligomer molecule. Therefore the differ-ence in the critical volume for the ion transport is small compared with that for

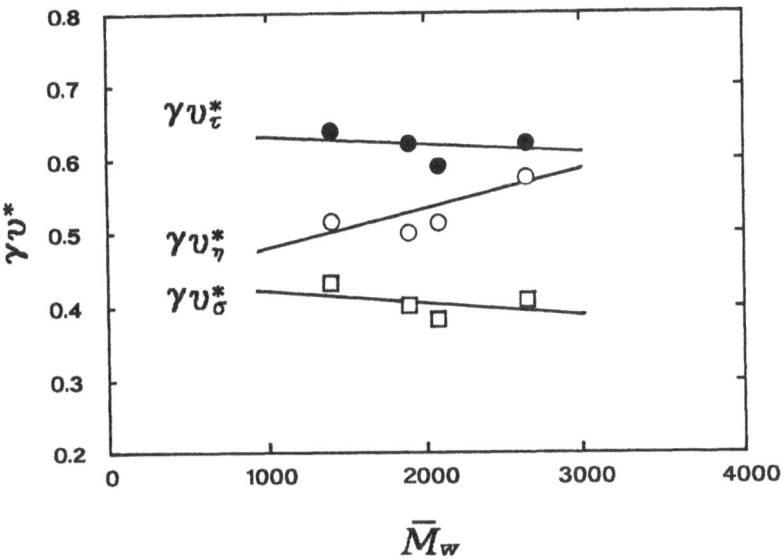

Fig. 19. Dependence of $\gamma\upsilon^*$ on $\bar{M}_w$ of DGEBA oligomer: $\bigcirc$ $\gamma\upsilon^*_\eta$ for segment mobility; $\bullet$ $\gamma\upsilon^*_\tau$ for dipole mobility; $\square$ $\gamma\upsilon^*_\sigma$ for ion mobility [113]

the segment transport among the DGEBA oligomers having different $\bar{M}_w$s from 1396 to 2640.

In contrast to the behavior of $\gamma\upsilon^*_\eta$ for the segment mobility, the $\gamma\upsilon^*_\tau$ value for the dipole mobility is not different among the oligomers and slightly decreases with an increase in the $\bar{M}_w$ of the oligomer as shown in Fig. 19. A larger $\bar{M}_w$ oligomer has a larger dipole and may need a larger critical volume for the transport of the moving unit, the dipole. The $\bar{M}_w$ dependent behavior of $\gamma\upsilon^*_\tau$ is difficult to explain by the size of the moving unit only. To understand the dipole mobility in the DGEBA oligomer, the mobility of the dipole will be compared with that of the chain segment.

8.2
Relationship Between Dipole and Chain Segment Mobility

The ratio of two $\gamma\upsilon^*$ values eliminates the undetermined factor γ and is a direct comparison of the mobility of two moving units in the same oligomer. The ratio of the dipole with the segment mobility, $\upsilon^*_\tau/\upsilon^*_\eta$, is plotted as a function of $\bar{M}_w$ for four DGEBA oligomers in Fig. 20 [113]. The mobility of the dipole is lower than that of the chain segment when the value of $\upsilon^*_\tau/\upsilon^*_\eta$ is greater than unity. Figure 20 indicates that the $\upsilon^*_\tau/\upsilon^*_\eta$ value is above unity and decreases with an increase in the $\bar{M}_w$ of the oligomer. The DGEBA oligomer having the $\bar{M}_w$ around 3000 has the $\upsilon^*_\tau/\upsilon^*_\eta$ value near unity. The mobility of the dipole is regarded to be smaller than the segment mobility in the DGEBA oligomers having

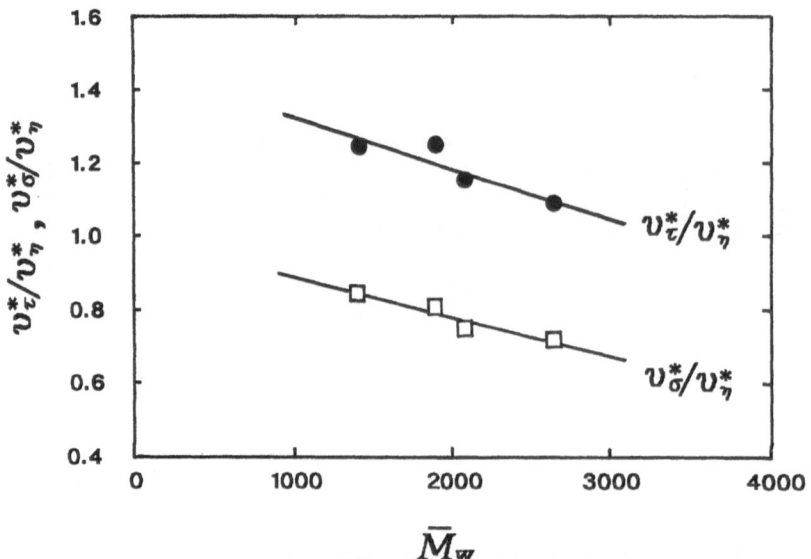

Fig. 20. Dependence of the ratio between critical volumes for two moving units on $\bar{M}_w$ of DGEBA oligomer: ● the ratio between critical volumes for dipole and segment mobility $(\upsilon^*_\tau/\upsilon^*_\eta)$; ❑ the ratio between critical volumes for ion and segment mobility $(\upsilon^*_\sigma/\upsilon^*_\eta)$ [113]

the $\bar{M}_w$ lower than 3000. A lower $\bar{M}_w$ oligomer has a greater $\upsilon^*_\tau/\upsilon^*_\eta$ value above unity as shown in Fig. 20, which suggests that the difference in mobility between the two moving units is relatively large in low $\bar{M}_w$ oligomers [113].

The dielectric relaxation time is defined as the time required for the polarization to decay to $1/e$ of the original equilibrium value. The dielectric relaxation time will become longer when the dipole movement is restricted by some force, such as molecular interaction. According to the chemical structure of the epoxide oligomer previously shown, each oligomer has a pair of strong polar end groups (epoxide ones) in the linear molecule. The molecular interaction, based on the terminal epoxide group, may produce a greater influence on the low $\bar{M}_w$ oligomer than on the high $\bar{M}_w$ one; the movement of a small-size dipole is more easily restricted than is that of a large-size dipole by this kind of molecular interaction [61, 113]. The influence of molecular interaction on the dipole mobility is considered small in higher $\bar{M}_w$ oligomers or polymers having larger-size dipoles.

The molecular interaction may also have influence on the segment mobility that is obtained from a macroscopic melt viscosity measurement for the epoxide oligomer. A low $\bar{M}_w$ DGEBA oligomer exhibits Newtonian flow behavior up to a relatively high shear rate, as previously shown. The zero shear melt viscosity of the DGEBA oligomer is usually obtained from the measurement under some degree of shear stress [61]. The influence of molecular interaction on the melt viscosity measurement under shear stress is considered small compared with the influence on the dielectric relaxation time measurement under a no-stress environment. A low $\bar{M}_w$ oligomer is affected more by the molecular interaction in the dielectric relaxation in comparison with a high $\bar{M}_w$ oligomer having larger dipoles. This may be an explanation for the $\bar{M}_w$ dependent of v^*_τ/v^*_η in Fig. 20. The mobility difference between the dipole and the chain segment is reflected in the viscoelastic behavior of the oligomer; the dielectric relaxation time is not always proportional to the melt viscosity for low $\bar{M}_w$ DGEBA oligomers [61, 113].

8.3
Relationship Between Ion and Chain Segment Mobility

The ionic conduction is closely related to the segment mobility of the host oligomer. The ratio of two critical volumes, v^*_τ/v^*_η, is a useful indication for the mobility difference between the two moving units, the ionic charge carrier and the chain segment. The v^*_σ/v^*_η value is plotted as a function of $\bar{M}_w$ for the DGEBA oligomers ($1396 \leq \bar{M}_w \leq 2640$) in Fig. 20 [113]. Each oligomer takes a v^*_σ/v^*_η value lower than unity, which indicates that the ionic charge carrier in the oligomer needs a smaller critical volume for the transport compared with the chain segment. Figure 20 also shows that the v^*_σ/v^*_η value decreases with an increase in the $\bar{M}_w$ of the oligomer. As already mentioned, each oligomer has similar kinds of ions that are far smaller than the chain segment. The size of the chain segment, on the other hand, increases with an increase in the $\bar{M}_w$ of the oligomer; the v^*_η value increases according to the increase of the $\bar{M}_w$ of the oligomer. Therefore, the ratio of the two critical volumes, v^*_σ/v^*_η, is below unity and decreases with an increase in the $\bar{M}_w$ of the oligomer. The v^*_σ/v^*_η value is about unity when extrapolated to zero $\bar{M}_w$ of the oligomer, which suggests that the mobility of the ion will be almost equal to that of a segment in a very small molecule. This may lead to the generally accepted concept, such as Walden's rule or $\sigma(T)\eta(T)=$const, for low molecular weight liquids.

9
Prediction of Tg of Curing System from Data on Epoxide Oligomers Before Crosslinking

9.1
Epoxy-Amine Curing System Before Gelation

The curing mechanism and the network formation behavior in epoxy-amine curing systems have been dealt with in several reviews of this volume, such as those by Dušek [129], and Rozenberg [130]. In the early and middle curing stages before gelation, the DGEBA-aromatic amine system may have a large proportion of linear molecules that are caused by the difference in the reaction rate between the primary and the secondary amine to the epoxy resin. Schechter et al.

[131] suggested that there were three possible reactions, [Eqs. (32) – (34)] in epoxy resins with diamines as follows:

$$R\text{-}NH_2 \;+\; CH_2\text{-}CH\text{--} \;\longrightarrow\; R\text{-}\overset{H}{N}\text{-}CH_2\text{-}CH\text{--} \qquad\qquad (32)$$
$$\underset{O}{\diagdown\diagup} \qquad\qquad\qquad \underset{OH}{}$$

$$R\text{-}\overset{H}{N}\text{-}CH_2\text{-}\underset{OH}{CH}\text{--} \;+\; CH_2\text{-}CH\text{--} \;\longrightarrow\; R\text{-}N \overset{\overset{OH}{|}}{\diagup CH_2\text{-}CH\text{--}} \qquad (33)$$
$$\underset{O}{\diagdown\diagup} \qquad\qquad \diagdown CH_2\text{-}\underset{OH}{CH}\text{--}$$

$$\underset{OH}{-CH\text{--}} \;+\; CH_2\text{-}CH\text{--} \;\longrightarrow\; \underset{O\text{-}CH_2\text{-}\underset{OH}{CH}\text{--}}{-CH\text{--}} \qquad\qquad (34)$$
$$\underset{O}{\diagdown\diagup}$$

Bell [132] reported that in the reaction between DGEBA and 4,4'-diamin-odiphenylmethane (DDM), the primary amine reaction [Eq. (32)] was approximately 7–12 times as fast as the secondary amine reaction [Eq. (33)] due to steric hindrance. The third reaction [Eq. (34)] is usually negligible. The DGEBA-DDM system before gelation may be a mixture of linear aromatic molecules with a small degree of branching. Therefore the uncured linear DGEBA oligomer is regarded as a simple model for the viscoelastic analysis of the reactive DGEBA-aromatic amine system before gelation.

It is beyond the scope of this review to discuss in detail viscoelastic properties at and after gelation. The gel point is one of the important characteristics in the epoxy resin curing process from both kinetic and rheological aspects. The gelation transition has been widely studied for epoxy resin systems [39, 44, 133–138] and already discussed in some of this series, such as by Malkin and Kulichikhin [28], Williams et al. [139], and Winter and Mours [140]. Rheological techniques for the determination of the gel point have been summarized for thermosetting resins by Halley et al. [29,141].

9.2
On-line Tg Prediction of Curing System

In the cure analysis of epoxy resin system, T_g is an important parameter for analyzing the curing behavior as described in a time-temperature-transformation (TTT) isothermal cure diagram that has been developed by Gillham et al. [18, 38, 40, 134, 142]. The T_g of the curing system, however, is difficult to measure directly during curing by the currently available measuring methods such as differential scanning calorimetry (DSC). Barton [143] has presented an overview for the application of DSC to the study of epoxy resin curing reactions. In contrast to the T_g measurement, on-line monitoring is possible in the melt viscosity, the dielec-

Table 13. On-line prediction method for glass transition temperature of epoxy resin curing system

Relationship	Remarks	Reference
A. $\sigma(T)[\eta(T)]^k=$ const $(0.63{\le}k{\le}1.12)$ $k=1.01-7.6\times10^{-3}\,T_g\ (-16{\le}T_g{\le}54\ ^\circ C)$	σ, ionic conductivity η; melt viscosity T; temperature k; exponent T_g; glass transition temperature	61, 144
B. $$\log\frac{\tau(T)}{42}=\frac{-C_1(T_g-T)}{C_2+T_g-T}$$	τ; dielectric relaxation time T; temperature C_1, C_2; constants T_g; glass transition temperature $(T_g>30\ ^\circ C)$	88, 147
C. $$\log\frac{\sigma(T)}{\sigma(T_g)}=\frac{C_1(T-T_g)}{C_2+T-T_g}$$ $C_2=C_3+C_4T_g, \log\sigma(T_g)=C_5+C_6T_g$	σ, ionic conductivity T; temperature $C_1, C_2, C_3, C_4, C_5, C_6$; constants[a] T_g; glass transition temperature	148, 149

[a] $C_1=10.5, C_2=109+0.58\,T_g$, and $\log\sigma(T_g)=-19.2+0.013\,T_g$ for DGEBA

tric relaxation time, and the ionic conduction measurements. The relationships between T_g and these three kinds of properties for uncured DGEBA oligomers are useful in the on-line T_g prediction of a curing system when the uncured DGEBA oligomer is regarded as a structural model for the DGEBA curing system.

Three methods have been proposed for the on-line T_g prediction of epoxy resin curing system as summarized in Table 13. These T_g prediction methods are based on the data of uncured DGEBA oligomers and essentially applicable to the curing stage before gelation. In the first method [61, 144], the T_g of the curing system is predicted from the following correlation between the melt viscosity and the ionic conductivity of the DGEBA oligomer before crosslinking:

$$\sigma(T)[\eta(T)]^k=\text{const and } k=1.01-7.6\times10^{-3}T_g \tag{35}$$

A simultaneous viscosity and ionic conduction measurement is necessary in this cure analysis. The k value in Eq. (35) has been reported to be constant around unity by some researchers, such as Tajima [97], Lane et al. [145], Stanford and McCullough [146], and Simpson and Bidstrup [81]. The exponent k is an indication of the mobility difference between the ionic charge carrier and the chain segment as previously discussed, so that k is not always constant for the reactive epoxy resin system increasing the molecular weight during curing. The variation in the exponent k is usually small for the epoxy resin system during the early stage of curing. A careful analysis of k in the DGEBA-DDM curing system reveals that the k value decreases with an advancement of reaction and is a useful

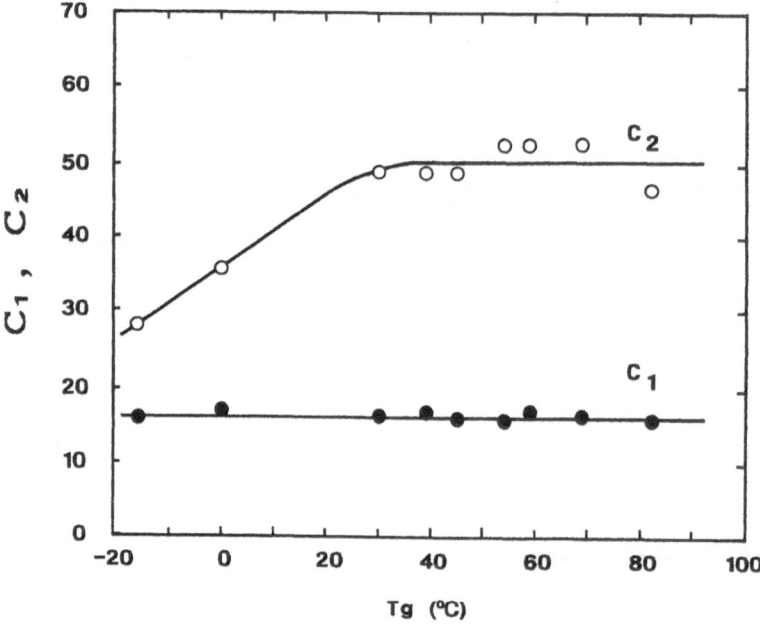

Fig. 21. WLF parameters, C_1 and C_2, of the dielectric relaxation time as a function of T_g for DGEBA oligomers: ● C_1; ○ C_2 [88]

indicator for predicting the T_g of the DGEBA-DDM system in the early curing stage [144].

The second method [88,147] in Table 13 is the application of a modified WLF equation describing the relationship between T_g and the dielectric relaxation time for DGEBA oligomers. In the modified WLF-type equation, T_g is regarded as a variable at a fixed temperature. The modified WLF type equation is derived from the experimental result that two WLF parameters and the τ value at T_g are similar, respectively, in some DGEBA oligomers. Figure 21 shows the T_g dependencies of two WLF parameters for the τ of DGEBA oligomer ($1396 \leq M_w \leq 9454$, $30 \leq T_g \leq 82$ °C). The two WLF parameters, C_1 and C_2, are almost universal values, respectively, for the DGEBA oligomers ($30 \leq T_g \leq 82$ °C). As previously mentioned, the τ value at T_g, $\tau(T_g)$, varies from 12 to 69 s for the same DGEBA oligomers discussed and is considered similar within experimental error [88]. Therefore T_g is regarded as a variable in the WLF equation for DGEBA oligomers ($T_g > 30$ °C) when the temperature T is fixed. Figure 22 demonstrates the WLF type relationships between the dielectric relaxation time and T_g at four fixed temperatures for DGEBA oligomers [88]. The T_g value of the curing system can be predicted through the modified WLF equation using the on-line dielectric measurement at each isothermal curing temperature. This prediction method is effective for the reactive DGEBA-curing agent system having T_g higher than 30 °C. A good agreement is reported between the calculated and actually measured T_gs for the

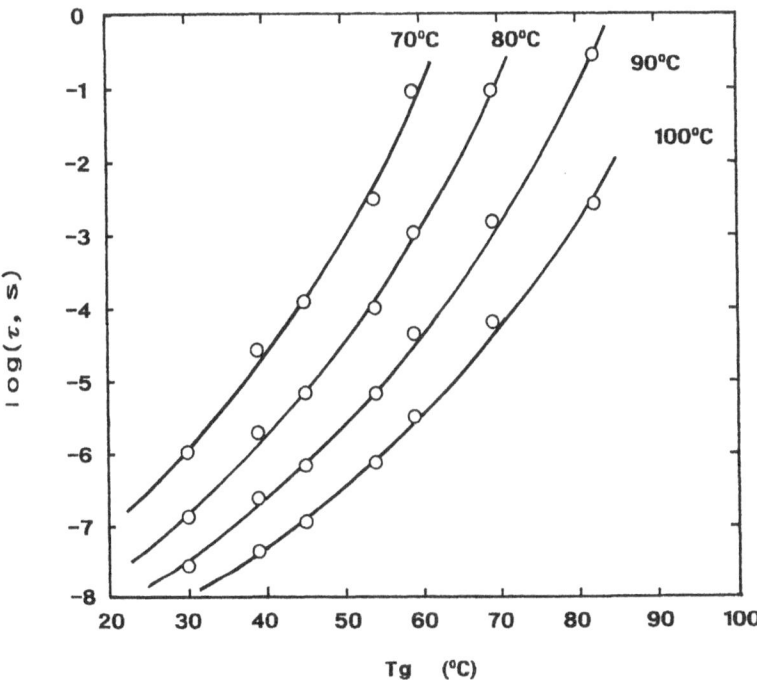

Fig. 22. Relationship between dielectric relaxation time and T_g for DGEBA oligomers at different temperatures [88]

DGEBA-DDM system during the intermediate curing stage before gelation [88, 147].

The third method [148,149] is based on another modified WLF equation modeling the relationship between T_g and the ionic conductivity for epoxide oligomers. Bidstrup et al. [148, 149] compared the modified WLF equation for the DGEBA-DDS curing system with that for the DGEBA oligomer before crosslinking. The ionic conductivity and T_g were modeled using the modified WLF type relationship not only in the DGEBA but also in the TGDDM based curing systems [148].

The T_g prediction methods listed in Table 13 have been mainly applied to the cure analysis of epoxy resin systems having a low degree of branching. A bisphenol-A novolac-type multifunctional epoxide oligomer, the poly(glycidyl ether) of bisphenol-A novolac (PGEBAN), has two side chains in every bisphenol-A backbone structure [61, 150]. The PGEBAN oligomer is regarded as a simple model for the DGEBA curing system having many branching. Viscoelastic data of the branched PGEBAN in combination with those of the non-branched DGEBA oligomer have been also used for the cure analysis of the DGEBA-bis (p-aminocyclohexyl) methane (PACM) system having some degree of branching before gelation [150].

10
Summary

Studies on three kinds of viscoelastic properties – melt viscosity, dielectric relaxation time, and ionic conductivity – have been reviewed for uncured DGEBA
oligomers above the T_g in this article. These viscoelastic properties can be correlated and discussed as the transport phenomena by considering the movements of three moving units: the chain segment, the dipole, and the ionic charge
carrier. Some empirical expressions have been proposed for describing the temperature dependencies of these three properties for DGEBA oligomers before
crosslinking. Each expression is based on the Williams-Landel-Ferry (WLF),
Vogel, Doolittle, or Berry & Fox type equation. As for the relaxation behavior,
the non-Debye model, such as the Havriliak-Negami or the Kohlraush-Williams-Watts (KWW) formula, governs the dielectric α-relaxation of the DGEBA
oligomer.
   The relationships between two of the three viscoelastic properties have been
reviewed and discussed for the DGEBA oligomer. Several new findings, which
appear to be different from polymers or simple liquids, have been reported for
DGEBA oligomers. One such example is that the dielectric relaxation time (τ) is
not always directly proportional to the melt viscosity (η). This phenomenon is
explained with the mobility difference between the dipole and the chain segment in the DGEBA oligomer; the mobility of the dipole is smaller than that of
the chain segment because the dipole movement is restricted by the molecular
interaction especially in the low molecular weight oligomer. The influence of the
molecular interaction is considered negligible in a higher molecular weight oligomer or a polymer having longer chain segment. This will be an interpretation
for the phenomenon widely believed; τ is directly proportional to η for a polymer. Other findings on the relationships between the three properties for the
DGEBA oligomer can be also explained with the mobility difference in the three
moving units: the chain segment, the dipole, and the ionic charge carrier.
   The same WLF-type equations are applicable to the temperature dependencies of the melt viscosity, the dielectric relaxation time, and the ionic conductivity for the uncured DGEBA oligomer. The two kinds of WLF parameters, C_1 and
C_2, have molecular weight dependencies in the DGEBA oligomer. The C_1 parameter is determined not only by the characteristics of an oligomer (a host oligomer) matrix but also by the size of the moving unit. The C_1 parameter was
found to be a good indicator for the mobility of each moving unit in the oligomer. The WLF C_2 parameter, on the other hand, reflects the characteristics of
an oligomer matrix or a host oligomer matrix only. Three WLF C_2 parameters,
obtained from the three viscoelastic properties in the same DGEBA oligomer,
take similar values and can be used to determine the fractional free-volume at
the T_g of the DGEBA oligomer. The free-volume fraction at the T_g is around 0.02
for the DGEBA oligomer and is a little smaller than the universal value (0.025).
   The uncured DGEBA oligomer is regarded as a simple structural model for
the reactive DGEBA-curing agent system before gelation. The viscoelastic prop-

erties of the uncured DGEBA oligomer can be used as the basic data for the cure analysis of the DGEBA-curing agent system at the point of the real-time monitoring. An oligomer has a molecular weight between a monomer and a polymer. Viscoelastic studies on epoxide oligomers before crosslinking provide useful information not only for the analysis of epoxy resin-curing agent systems but also for a better understanding of the nature of the polymer or monomer from a different point of view.

References

1. Lee H, Neville K (1967) Handbook of Epoxy Resins. McGraw-Hill, New York
2. May CA (1988) Epoxy Resins, Chemistry and Technology, 2nd edn. Marcel Dekker, New York
3. McAdams LV, Cannon JA (1986) In: Kroschwitz JI (ed) Encyclopedia of Polymer Science and Engineering 2nd Ed Vol 6. Wiley-Interscience, New York, p 322
4. Batzer H, Zahir SA (1975) J Appl Polym Sci 19:601
5. Bantle S, Burchard W (1986) Polymer 27:728
6. Eppert G, Liebscher G (1982) J Chromatogr 238:399
7. Mori S (1981) Anal Chem 53:1813
8. Hata N, Kumanotani J (1973) J Appl Polym Sci 17:3545
9. Scheuing DR (1985) J Coatings Technol 57(723):47
10. Koike T, Tanaka R (1991) J Appl Polym Sci 42:1333
11. Macho E, Alegría A, Colmenero J (1987) Polymer 27:810
12. Alvarez F, Alegría A, Colmenero J (1991) Phys Rev B 44:7306
13. Alvarez F, Alegría A, Colmenero J (1993) Phys Rev B 47:125
14. Alegría A, Guerrica-Echevarría E, Tellería I, Colmenero J (1993) Phys Rev B 47:14857
15. Colmenero J, Arbe A, Alegría A (1994) J Non-cryst Solids 172–174:126
16. Mertzel E, Koenig JL (1986) Adv Polym Sci 75:73
17. Stevens GC (1981) J Appl Polym Sci 26:4279
18. Wang X, Gillham JK (1991) J Appl Polym Sci 43:2267
19. Sojka SA, Moniz WB (1976) J Appl Polym Sci 20:1977
20. Barton JM, Buist GJ, Howlin BJ, Jones JR, Liu S, Hamerton I (1994) Polym Bull 33:215
21. Ochi M, Tanaka Y, Shimbo M (1975) Nippon Kagaku Kaishi 1975:1600
22. Dušek K, Bleha M, Luňák S (1977) J Polym Sci, Polym Chem Ed 15:2393
23. Shimbo M, Ochi M, Nakaya T (1982) Nippon Kagaku Kaishi 1982:1400
24. Horie K, Hiura H, Sawada M, Mita I, Kambe H (1970) J Polym Sci, Part A-1 8:1357
25. Peyser P, Bascom WD (1977) J Appl Polym Sci 21:2359
26. Tajima YA, Crozier D (1983) Polym Eng Sci 23:186
27. Mijovic J, Lee CH (1989) J Appl Polym Sci 37:889
28. Malkin AY, Kulichikhin SG (1991) Adv Polym Sci 101:217
29. Halley PJ, Mackay ME (1996) Polym Eng Sci 36:593
30. Goto K, Kawai R (1978) Kobunshi Ronbunshu 35:75
31. Malkin AY, Zhirkov PV (1990) Adv Polym Sci 95:111
32. Suzuki K, Miyano Y, Kunio T (1977) J Appl Polym Sci 21:3367
33. Strehmel V, Fryauf K, Sommer C, Arndt KF, Fedtke M (1992) Angew Makromol Chem 196:195
34. Ciercierski G, Kohman Z (1981) Int Polym Sci Technol 8:T-105
35. Booss HJ, Hauschildt KR (1980) Kunststoffe 70:48
36. Adolf D, Chambers R (1997) Polymer 38:5481
37. Naganuma T, Sakurai T, Takahashi Y, Takahashi S (1972) Kobunshi Kagaku 29:105
38. Babayevsky PG, Gillham JK (1973) J Appl Polym Sci 17:2067
39. Lairez D, Adam M, Emery JR, Durand D (1992) Macromolecules 25:286

40. Wisanrakkit G, Gillham JK (1990) J Appl Polym Sci 41:2885
41. Pascault JP, Williams RJJ (1990) J Polym Sci, Part B Polym Phys 28:85
42. Lane JW, Seferis JC, Bachmann MA (1986) Polym Eng Sci 26:346
43. Wu S, Gedeon S, Fouracre RA (1988) IEEE Trans Electr Insul 23:409
44. Parthun MG, Johari GP (1992) Macromolecules 25:3254
45. Deng Y, Martin GC (1994) Macromolecules 27:5147
46. Adamec V (1972) J Polym Sci, Part A-1 10:1277
47. Martin BG (1977) Mater Eval 35 (6):48
48. Kranbuehl D, Delos S, Hoff M, Weller L, Haverty P, Seeley J (1988) ACS Symp Ser 367:100
49. Ulañski J, Friedrich K, Boiteux G, Seytre G (1997) J Appl Polym Sci 65:1143
50. Saksena TK, Babbar NK (1979) Ultrasonics 17:122
51. Bunton LG, Daly JH, Maxwell ID, Pethrick RA (1982) J Appl Polym Sci 27:4283
52. Suzuki T, Oki Y, Numajiri M, Miura T, Kondo K, Ito Y (1993) Polymer 34:1361
53. Levy RL, Ames DP (1984) Polym Sci Technol 29:245
54. Wang FW, Lowry RE, Fanconi BM (1986) Polymer 27:1529
55. Billingham NC, Calvert PD, Ghaemy MG (1983) Br Polym J 15:62
56. Mackinnon AJ, Pethrick RA, Jenkins SD, McGrail PT (1992) Macromolecules 25:3492
57. Aleman JV (1978) Polym Eng Sci 18:1160
58. Aleman JV (1980) J Polym Sci, Polym Chem Ed 18:2567
59. Ghijsels A, Groesbeek N, Raadsen J (1984) Polymer 25:463
60. Utracki LA, Ghijsels A (1987) Adv Polym Technol 7:35
61. Koike T (1993) J Appl Polym Sci 47:387
62. Simpson JO, Bidstrup SA (1993) J Polym Sci, Part B Polym phys Ed 31:609
63. Doolittle AK (1951) J Appl Phys 22:1471
64. Vogel H (1921) Physik Z 22:645
65. Fulcher GA (1925) J Am Ceram Soc 8:339
66. Williams ML, Landel RF, Ferry JD (1955) J Am Chem Soc 77:3701
67. Berry GC, Fox TG (1968) Adv Polym Sci 5:261
68. Marwedel G (1981) Chem Ztg 105:79
69. Hadjistamov D (1990) Farbe Lack 96:15
70. Lin SC, Pearce EM (1979) J Polym Sci, Polym Chem Ed 17:3095
71. Delmonte J (1959) J Appl Polym Sci 2:108
72. Zukas WX, MacKnight WJ, Schneider NS (1983) ACS Symp Ser 227:223
73. Lane JW, Seferis JC, Bachmann MA (1986) J Appl Polym Sci 31:1155
74. Nass KA, Seferis JC (1989) Polym Eng Sci 29:315
75. Wetton RE, Foster GM, Gearing JWE, de Blok M (1989) J Therm Anal 35:469
76. Mangion MBM, Johari GP (1990) J Polym Sci, Part B Polym Phys 28:1621
77. Mangion MBM, Johari GP (1991) J Polym Sci, Part B Polym Phys 29:437
78. Senturia SD, Sheppard Jr NF, Lee HL, Day DR (1982) J Adhesion 15:69
79. Senturia SD, Sheppard Jr NF (1986) Adv Polym Sci 1:80
80. Gotro J, Yandrasits M (1989) Polym Eng Sci 29:278
81. Simpson JO, Bidstrup SA (1995) J Polym Sci, Part B Polym Phys 33:55
82. Kranbuehl D, Delos S, Hoff M, Haverty P, Freeman W, Hoffman R, Godfrey J (1989) Polym Eng Sci 29:285
83. Debye P (1929) Polar Molecules. Chemical Catalog, New York
84. Cole KS, Cole RH (1941) J Chem Phys 9:341
85. Davidson DW, Cole RH (1950) J Chem Phys 18:1417
86. Havriliak S, Negami S (1966) J Polym Sci, Part C 14:99
87. Williams G, Watts DC (1970) Trans Faraday Soc 66:80
88. Koike T (1992) J Appl Polym Sci 45:901
89. Nass KA, Seferis JC (1990) Thermochimica Acta 170:19
90. McCrum NG, Read BE, Williams G (1967) Anerastic and Dielectric Effects in Polymeric Solids. John Wiley & Sons, London

91. Takeishi S, Mashimo S (1982) Rev Sci Instrum 53:1155
92. Koike T (1993) Polym Eng Sci 33:1301
93. Sheppard Jr NF, Senturia SD (1989) J Polym Sci, Part B Polym Phys 27:753
94. Paluch M, Ziolo J, Rzoska SJ (1997) Phys Rev B 56:5764
95. Saito S, Sasabe H, Nakajima T, Yada K (1968) J Polym Sci, Part A–6:1297
96. Walden P (1906) Z Phys Chem 55:207
97. Tajima YA (1982) Polym Comp 3:162
98. Sazhin BI, Shuvayev VP (1965) Vysokomol soyed 7:962
99. Koike T (1995) J Appl Polym Sci 56:1183
100. Strobl GR (1996) The Physics of Polymers. Springer, Berlin Heidelberg New York
101. Ferry JD (1970) Viscoelastic Properties of Polymers. John Wiley, New York
102. Sasabe H, Saito S (1972) Polymer J 3:624
103. Birks JB, Hart J (1961) Progress in Dielectrics. Heywood & Company, London
104. Hill NE, Vaughan WE, Price AH, Davies M (1969) Dielectric Properties and Molecular Behaviour. Van Nostrand Reinhold, London
105. Cohen MH, Turnbull D (1959) J Chem Phys 31:1164
106. Turnbull D, Cohen MH (1961) J Chem Phys 34:120
107. Jean YC, Sandreczki TC, Ames DP (1986) J Polym Sci, Part B Polym Phys 24:1247
108. Deng Q, Sundar CS, Jean YC (1992) J Phys Chem 96:492
109. Deng Q, Zandiehnadem F, Jean YC (1992) Macromolecules 25:1090
110. Deng Q, Jean YC (1993) Macromolecules 26:30
111. Wang YY, Nakanishi H, Jean YC, Sandreczki TC (1990) J Polym Sci, Part B Polym Phys 28:1431
112. Suzuki T, Oki Y, Numajiri M, Miura T, Kondo K, Shiomi Y, Ito Y (1993) J Appl Polym Sci 49:1921
113. Koike T, Ishizaki N (1999) J Appl Polym Sci 71:207
114. Simha R, Boyer RF (1962) J Chem Phys 37:1003
115. Boyer RF, Simha R (1973) J Polym Sci, Polym Lett Ed 11:33
116. Plazek DJ, Choy IC (1989) J Polym Sci, Part B Polym Phys 27:307
117. Shito N (1968) J Polym Sci, Part C 23:569
118. Miyamoto T, Shibayama K (1972) Kobunshi Kagaku 29:467
119. Gupta VB, Brahatheeswaran C (1991) Polymer 32:1875
120. Soh R, Yamamoto S, Shimbo M (1981) Nippon Setchaku Kyokaishi 17:507
121. Shigeta Y, Ochi M, Shimbo M (1981) J Appl Polym Sci 26:2256
122. Bellenger V, Dhaoui W, Morel E, Verdu J (1988) J Appl Polym Sci 35:563
123. Ogata M, Kinjyo N, Kawata T (1993) J Appl Polym Sci 48:583
124. Miyamoto T, Sugano T, Shibayama K (1973) Kobunshi Kagaku 30:155
125. Oleinik EF (1986) Adv Polym Sci 80:49
126. Bellenger V, Dhaoui W, Verdu J (1987) J Appl Polym Sci 33:2647
127. Gupta VB, Brahatheeswaran C (1989) J Appl Polym Sci 38:1957
128. Gupta VB, Brahatheeswaran C (1990) J Appl Polym Sci 41:2533
129. Dušek K (1986) Adv Polym Sci 78:1
130. Rozenberg BA (1986) Adv Polym Sci 75:113
131. Schechter L, Wynstra J, Kurkjy RP (1956) Ind Eng Chem 48:94
132. Bell JP (1970) J Polym Sci, Part A–6:417
133. Tung CYM, Dynes PJ (1982) J Appl Polym Sci 27:269
134. Enns JB, Gillham JK (1983) J Appl Polym Sci 28:2567
135. Winter HH (1987) Polym Eng Sci 27:1698
136. Heise MS, Martin GC, Gotro JT (1992) Polym Eng Sci 32:529
137. Lairez D, Emery JR, Durand D, Pethrick RA (1992) Macromolecules 25:7208
138. Mijovic J, Kenny JM, Nicolais L (1993) Polymer 34:207
139. Williams RJJ, Rozenberg BA, Pascault JP (1997) Adv Polym Sci 128:95
140. Winter HH, Mours M (1997) Adv Polym Sci 134:165
141. Halley PJ, Mackay ME, George GA (1994) High Perform Polym 6:405

142. Aronhime MT, Gillham JK (1986) Adv Polym Sci 78:83
143. Barton JM (1985) Adv Polym Sci 72:111
144. Koike T (1993) J Appl Polym Sci 50:1943
145. Lane JW, Khattak RK, Dusi MR (1989) Polym Eng Sci 29:339
146. Sanford WM, McCullough RL (1990) J Polym Sci, Part B Polym Phys 28:973
147. Koike T (1992) J Appl Polym Sci 44:679
148. Bidstrup WW, Senturia SD (1989) Polym Eng Sci 29:290
149. Bidstrup SA, Sheppard Jr NF, Senturia SD (1989) Polym Eng Sci 29:325
150. Koike T, Ishizaki N (1998) Polym Eng Sci 38: 1838

Editor: Prof. K. Dusek
Received: February 1999

Author Index Volumes 101–148

Author Index Volumes 1–100 see Volume 100

Subject Index

Springer
and the
environment

At Springer we firmly believe that an
international science publisher has a
special obligation to the environment,
and our corporate policies consistently
reflect this conviction.

We also expect our business partners –
paper mills, printers, packaging
manufacturers, etc. – to commit
themselves to using materials and
production processes that do not harm
the environment. The paper in this
book is made from low- or no-chlorine
pulp and is acid free, in conformance
with international standards for paper
permanency.

 Springer